Mathematics Education in the United States 2016

A Capsule Summary Fact Book

Written for
The Thirteenth International Congress
on Mathematical Education (ICME-13)

Hamburg, Germany, July 2016

by
John A. Dossey
Sharon Soucy McCrone
Katherine Taylor Halvorsen

under the Auspices of the
National Council of Teachers of Mathematics
and the
United States National Commission on Mathematics Instruction

The preparation, production, and dissemination of this document were funded by National Science Foundation under grant number DRK-12; DRL 1503277; *Thirteenth International Congress on Mathematical Education (ICME-13) Travel Grant* to the National Council of Teachers of Mathematics, Gail Burrill, Principal Investigator. The opinions expressed herein are those of the authors and do not necessarily reflect the views of the National Science Foundation, the National Council of Teachers of Mathematics, or the U.S. National Commission on Mathematics Instruction.

Copyright © 2016 by
THE NATIONAL COUNCIL OF TEACHERS OF MATHEMATICS, INC.
1906 Association Drive, Reston, VA 20191-1502
(703) 620-9840; (800) 235-7566; www.nctm.org

Library of Congress Cataloging-in-Publication Data

Names: Dossey, John A. | McCrone, Sharon. | Halvorsen, Katherine. |
  International Congress on Mathematical Education (13th : 2016 : Hamburg,
  Germany) | National Council of Teachers of Mathematics. | United States
  National Commission on Mathematics Instruction.
Title: Mathematics education in the United States 2016 : a capsule summary
  fact book : written for the Thirteenth International Congress on
  Mathematical Education (ICME-13), Hamburg, Germany, July 24–31, 2016 / by
  John A. Dossey, Sharon Soucy McCrone, Katherine Taylor Halvorsen ; under
  the auspices of the National Council of Teachers of Mathematics, and the
  United States National Commission on Mathematics Instruction.
Description: Reston, VA : National Council of Teachers of Mathematics, [2016]
  | Includes bibliographical references.
Identifiers: LCCN 2016001208 | ISBN 9780873539821 (pbk.)
Subjects: LCSH: Mathematics—Study and teaching—United States.
Classification: LCC QA13 .D724 2016 | DDC 510.71/073—dc23 LC record available at
  http://lccn.loc.gov/2016001208

The National Council of Teachers of Mathematics is the public voice of mathematics education, supporting teachers to ensure equitable mathematics learning of the highest quality for all students through vision, leadership, professional development, and research.

Printed in the United States of America

Contents

Contents—*Continued*

Preface

The population of the United States is approximately 323,996,000. Approximately 18% of these individuals are formally enrolled in a public or private elementary or secondary school or are homeschooled, while nearly another 7% are enrolled as students in a degree-granting postsecondary institution. In the entire U.S. population in 2014, about 66% were 25 years or older, and of those adults, 88% had completed high school or its equivalent, and about 30% had at least a bachelor's degree (Snyder and Dillow 2011; U.S. Department of Commerce 2014).

No single government agency controls public education in K–grade 12 in the United States. Rather, authority for most educational decisions lies with education agencies in the 50 individual states, which in turn share decision making with the individual school districts within their borders. In the 2015–16 academic year, U.S. public schools educated approximately 50,773,000 students, private elementary and secondary schools contributed another 5,183,000 students, and homeschooling accounted for another 1,775,000 students (Snyder and Dillow 2015). Similarly, both public and private institutions exist at the college and university level, with authority for state institutions residing at a mixture of state and local levels for public institutions and at the institutional level for most private institutions. In the 2012–13 academic year, 4,726 accredited institutions offered degrees at the associate's level or above. These included 1,623 public institutions, 1,652 private not-for-profit institutions, and 1,451 private for-profit institutions. Of the total of 4,726 institutions, 3,026 awarded degrees at the bachelor's level or higher, and 1,700 offered associate's degrees as their highest degree (Snyder and Dillow 2015).

Determining what is happening in such a large and complex country as the United States is quite difficult, even for those in the United States and others who are familiar with U.S. education. Many at conferences of the International Congress on Mathematical Education (ICME) are unfamiliar with education in the United States. Consequently, in 1999, the U.S. National Commission on Mathematics Instruction recommended that the National Council of Teachers of Mathematics (NCTM) request funds from the National Science Foundation (NSF) to bring together available data

about mathematics education in the United States for a document to be distributed at the Ninth International Congress on Mathematical Education (ICME-9), held in 2000, to provide mathematics educators throughout the world with information about this complex system. This process was repeated for subsequent ICMEs, held in 2004, 2008, and 2012. The present publication now extends the series with information available as of the end of 2015.

This report begins with some general information about education in the United States. It then describes the three kinds of curricula identified in the Second International Mathematics Study—intended, implemented, and attained (McKnight et al. 1987)—and gives special attention to the emergence of a common K–grade 12 curriculum that has been adopted by 43 states and the District of Columbia. This curriculum, the Common Core State Standards for Mathematics (CCSSM), was developed by a consortium consisting of state governors and chief state education officers (National Governors Association Center for Best Practices and Council of Chief State School Officers [NGA Center and CCSSO] 2010a, 2010b). The adoption of such a set of common outcomes, matching assessments, and similar instructional materials is expected to bring a new level of uniformity and coherence to U.S. mathematics education. This report examines the current state of adoption, adaptation, and implementation of those new standards and surveys the resulting shifts in expectations, content, instruction, and learning opportunities emanating from attempts to align school mathematics programs with them. The report also examines the Every Student Succeeds Act, the federal education law passed by the U.S. Congress at the end of 2015 to revise, update, and reauthorize the No Child Left Behind Act, which had defined the U.S. government's role in public elementary and secondary education from 2002 to that time.

This report on mathematics education in the United States consists of nine chapters. A brief survey of their focus and content may help readers orient themselves and navigate through them.

Chapter 1 presents a general overview of public and private educational opportunities in the United States, including the movement of U.S. students through the K–12 school years and on to admission to postsecondary education. It also looks at the recent passage of the Every Student Succeeds Act of 2015.

Chapter 2 gives an overview of the history and current status of the intended curriculum for school mathematics—its origins and goals. This portion of the report gives a listing and discussion of the documents and movements that have given U.S. mathematics education its current shape and have influenced the forces that are currently acting on it.

Chapter 3 examines what is known about the actual implemented K–12 curriculum, instructional approaches, and materials in use and considers evolving changes in the postsecondary curricula.

Growing naturally out of Chapter 3, Chapter 4 addresses the attained curriculum. It examines the extant outcomes from national and international assessments of student achievements in mathematics and problem solving. The national assessments survey state- and national-level performances on the National Assessment of Educational Progress (NAEP). Chapter 4 ends with an examination of student performance on college entrance examinations.

In Chapter 5, the focus is on U.S. student achievement outcomes in international comparative studies—Trends in International Mathematics and Science Study (TIMSS) and the Programme for International Student Assessment (PISA). Such results give a glimpse of how the performance of U.S. students compares with that of their international peers and provide a basis for asking questions about the impact of various factors in education and social environments that may offer explanations for differences in international student achievement in mathematics.

The remaining chapters of the book examine the following:

- Chapter 6: The Common Core State Standards for Mathematics (NGA Center and CCSSO 2010a, 2010b), with an emphasis on current professional development efforts and resources for teachers

- Chapter 7: Changes in school formats (charter, private religious, private nonreligious, and home-schooling) and their challenges for transitions between levels of schooling; mathematics teacher education and professional development programs; and new resources for teachers and professional development focusing on data and statistics and on weaving them into school curricula

- Chapter 8: Special programs for accelerated students at the K–12 school and postsecondary levels, as well as national and international competitions in mathematics for students

- Chapter 9: Professional organizations and resources for teachers of mathematics

One message that comes through repeatedly in this report and its descriptions is that the variety of education programs available in the United States is very great, and thus the possibility of characterizing them adequately in a brief document like this one is very small. Another message is that all levels of the U.S. educational system exhibit great flux, and even though we have attempted to provide the latest information available, we realize that the content that we present in this report will quickly become dated. By listing our sources, we hope to enable interested readers to obtain updated information.

Finally, we would like to acknowledge the efforts of Gail Burrill, who wrote the proposal for the grant under which the funding for this publication was obtained, as well as the insightful, constructive, and editorially valuable advice that Solomon Friedberg, Matt Larson, Roxy Peck, James Roznowski, and John Staley provided during the development of the report, and the fine work of Anita Draper and Rebecca Totten at NCTM in editing and producing this document. We have tried to be as accurate as possible and apologize for any errors.

Chapter I: General Information on U.S. Education

We begin with a general overview of public and private educational opportunities in the United States. This discussion will provide a background for our subsequent, more detailed examination of mathematics education in the United States in 2016. The final portion of the chapter devotes attention to the new federal education law in the United States, the Every Student Succeeds Act, passed at the end of 2015.

Overall Organization of Education in the United States

Figure 1 presents a graphical overview of the structure of education in the United States. The system can be thought of as consisting of four broadly defined levels: elementary school (K–grade 5 or K–grade 6, corresponding to ages 5–10 or 11); middle school or junior high school (grades 6–8 or 7–8, ages 11–13 or 12–13, respectively); senior high school (grades 9–12, ages 14–17); and postsecondary, or tertiary, education (grades 13 and above, ages 18 and older). The ending and beginning points of the each of the levels varies, owing to state and local school system regulations and preferences (Snyder and Dillow 2015).

The numbered scales up the margins of figure 1 indicate, on the left-hand side, the median ages for students enrolled at the varied levels of K–12 education and, on the right-hand side, the corresponding levels from pre-kindergarten through grade 12 of education and the years normally taken for a full-time student to progress through the varied levels of tertiary education. One can loosely interpret the width of the horizontal bars associated with school organizations at different ages as representing the percentage of students enrolled in the varied forms of education at the K–12 levels. Later, additional commentary will amplify the impact of students "dropping out" of education before completion of grade 12 or leaving education to join the workforce after completing grade 12. At the community college level, one must also understand that community and junior colleges may provide vocational and technical education programs for students.

Movement of U.S. Students through K–12 Education

K–12 students are legally required to start and maintain enrollment in formal education by state-mandated ages. The minimum compulsory school-starting ages range from 5 to 8 years (age 5 [8 states], age 6 [25 states], age 7 [15 states], and age 8 [2 states]). Standards for the length of compulsory education also vary by state, with minimum allowed school-leaving ages of 16 to 18 (age 16 [23 states], age 17 [9 states], and age 18 [18 states]). Eight states simply require 9 years of formal education, while 4 states require a total of 13 years. However, state standards in nearly half of the 50 states allow for variances in their regulations for school-starting and school-leaving ages for students who are employed; have a physical or mental condition that makes attendance infeasible; have passed eighth grade successfully; or have the permission of their parents, district court, or school board (Bush 2010, Mikulecky 2013). The variance in these regulations across the 50 states is mirrored by the diversity in laws respecting when schooling should begin and what constitutes the minimum amount of schooling acceptable for students in a state. Another example of diversity in education across the states manifests itself in the variability of the NAEP achievement results reported in table 9 in Chapter 4. These two examples reflect differences in state standards, state expectations for students, and the structure of state funding.

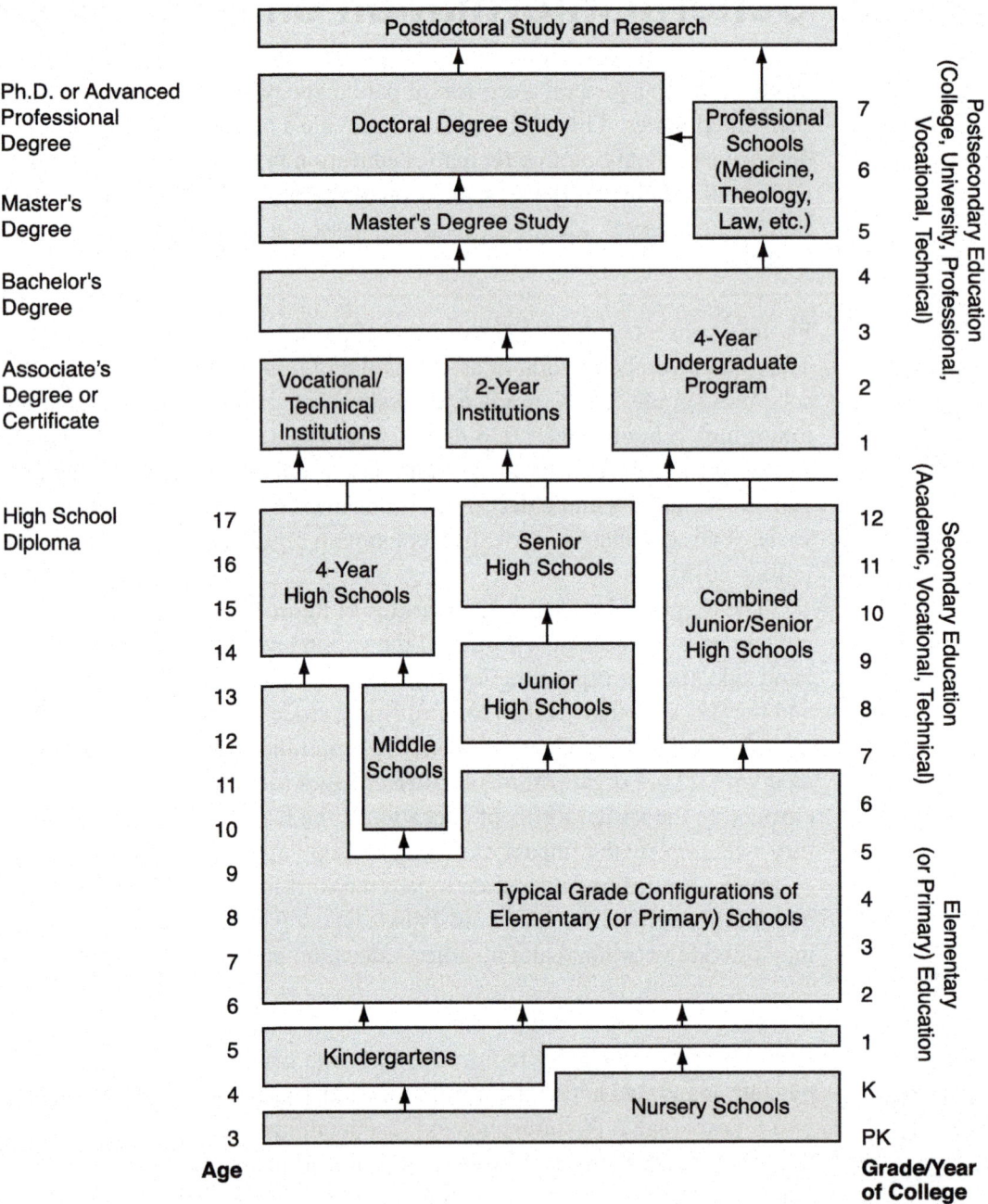

Fig. 1. The structure of education in the United States (Snyder and Dillow 2015)

Not all U.S. students complete secondary education prior to leaving formal education. Although state laws require compulsory education, they also allow for homeschooling of students by their parents. The percentage of students who complete a public school education can be quantified in many ways (Stetser and Stillwell 2014). The *average freshman graduation rate* (AFGR) provides an estimate of the proportion of public high school students who graduate from high school four years after having entered the ninth grade. Using the AFGR and the Common Core of Data compiled by the National Center of Educational Statistics, NCES statisticians have been able to develop a trend line for this measure of student persistence and completion. Of those who entered high school as ninth graders in the academic year 2008–9, the files suggest that 81% finished during the 2011–12 school year. This was the highest completion rate in a four-year period since the trend line was first developed for *The Condition of Education* in 1990–91, a congressionally mandated annual report to Congress describing the current status of education K–college in the United States. In the inaugural year, the AFGR was 74%. The statistic dropped to 71% in 1995–96 and stayed at that level in 1998–99. It then increased to 75 % in 2004–5, dropped again to 73% in 2005–6, and then increased to 80% in 2010–11 before increasing again to the 81% mentioned above for those graduating in 2011–12. Among the 2011–12 graduates, the AFGR rates for various racial/ethnic groups were as follows: Asian/Pacific Islanders (93%), White (85%), Hispanic (76%), and both Black and American Indian/Alaskan Native (68%) (Kena et al. 2015). Students who do not complete high school with their class in four years may continue their enrollment until receiving their diplomas later.

The many students who discontinue their education may achieve the equivalent of a high school diploma through other means. The *status completion rate* (SCR), another completion ratio, provides the percentage of people by age ranges who are not attending a secondary school but have earned a high school diploma or have completed a high school equivalency program. In the 18- to 24-year-old age group, the SCR in 2008 was 89.9%, compared with 87.2% in 2002 and 83.9% in 1980. Gender comparisons for 2008 showed that 90.5% of females and 89.3% of males had achieved a high school diploma or its equivalent, but major differences exist among racial or ethnic subgroups: 94.2% for White non-Hispanics students, 86.9% for Black non-Hispanic students, and 75.5% for Hispanics (Chapman, Laird, and Kewal-Ramani 2010; U.S. Department of Commerce, Census Bureau, American Community Survey 2007, 2012).

The government examined the SCR again in 2012 and found that it had increased to 91.3% with gender-based status completion rate comparisons of male (90.3%) and female (92.3%). The corresponding 2012 SCR figures for cultural ethnic/racial groups are as follows: White non-Hispanics, 94.6%; Black non-Hispanics, 90.0%; and Hispanics, 87.2% (Stark and Noel 2015). The upward trend in overall SCR improvement that began in 1980 continued into 2012. In other 2012 comparisons, females ages 18–24 had a higher SCR than males, and White non-Hispanic students had a higher SCR than Black-non-Hispanic students, who in turn had a higher SCR than Hispanic students. Although the overall SCR for 18- to 24-year-olds is encouraging, the large gaps in percentages of completers by racial or ethnic groups provide a challenge to those involved in U.S. secondary education and literacy programs (Stark and Noel 2015). These data reflect the fact that the United States is a nation of immigrants, bringing their languages to urban, suburban, small town, and rural settings. Often they are met with cultural and economic challenges that impede their opportunities to progress through the U.S. education system. The analysis of completion data is one way of seeing whether all students are finally enjoying opportunities to succeed in securing the equivalent of a high school diploma.

Movement of U.S. Students through Postsecondary Education	Students who graduate from high school may enter the workforce, attend a non-university tertiary institution focusing on technical or vocational education, attend a two-year community college, or attend a four-year college or university. At this level, the bars in figure 1 represent the flow of students still in the educational stream. Two-year and community colleges usually offer diverse selections of courses and programs, including those that overlap with the first two years of the curriculum at a four-year college, along

with a number of courses that overlap with those found in the technical colleges and high schools. Many community colleges also have vocational streams of students who earn certification for a particular career, sometimes with and sometimes without a two-year degree.

In two-year or community colleges, an associate of arts (AA), an associate of sciences (AS), or an associate of applied sciences (AAS) degree can usually be earned through the equivalent of two years of full-time study. One-year certificate programs are also offered in various technical fields. In addition, a number of vocational or trade schools offer programs in which students can focus on the knowledge and skills needed to perform a particular job. Vocational schools may be integrated with public schools as part of programs that facilitate the transition from school to work. In other instances, these schools are private schools, nonprofit or proprietary, operated outside the public school system. The foci of these schools range from apprenticeship programs for trades to culinary institutes.

U.S. four-year colleges and universities offer bachelor of science (BS) and bachelor of arts (BA) degrees that can typically be completed in four years of full-time study. In addition, many universities offer graduate programs leading to master's (MS, MA, or MEd) degrees and doctoral (PhD and EdD) degrees. Programs leading to professional degrees (law, medicine, business, etc.) exist both in universities and in institutions that offer no other degree programs. The time needed to complete post-bachelor degrees varies with the field and institution.

The U.S. Education Enterprise	In 2013–14, 98,271 public schools or agencies were in operation in the 50 states and the District of Columbia. These schools were providing a variety of educational services to the estimated 50 million K–12 students enrolled in them. Most of the schools (89,183) were focused on delivering the broad standard curriculum to their students. Another 1,380 provided targeted vocational or technical education, while 2,010 offered special education services. Another 5,986 offered some form of alternative education. Included in this number were 2,779 independent charter agencies (not including those already counted because they are imbedded directly in the curriculum of a public school program). These operational schools were part of one of 18,184 operational public school districts in the United States, ranging from 1,252 districts in Texas to 19 districts in Delaware and excluding Hawaii, which is a single-district state. In 2012–13, these districts employed the equivalent of 3.1 million full-time teachers (Glander 2015).

In addition to the public schools that are created by local communities and provide education to that community's youth in accordance with state regulations, other types of public, private, and home schools operate at the K–12 levels in the United States. Charter schools are public schools that are funded through public and state support but are allowed to operate with freedom from many of the regulations that apply to traditional public schools. Magnet schools are public schools whose curricula address the standard requirements and regulations but provide targeted and advanced instruction in such areas as mathematics, science, or the arts. The provisions governing magnet schools also usually include a requirement that specific percentages of students come from particular cultural, ethnic, or racial groups in a school's or district's student body.

In addition to these public- and state-funded schools, many communities also have private schools that are funded by parents of the students enrolled, religious denominations, community foundations, or other donors. These schools are independent, nongovernmental, or non-state, schools. The school's administration is typically responsible to a council or board, often established by the parents of the students attending. Public and magnet schools' administrators are commonly responsible to a governing board elected by the public of the geographical area that the school serves. Charter schools' administrators are typically responsible to a board elected by the parents of the students, and they also are accountable to varied local and state regulations, depending on the laws of the state in which they are located.

Private schools are funded primarily by tuition charged annually to students' parents. Some scholarship aid may be available through a school foundation or established by past graduates. Annual tuition for private schools ranges from nothing at schools whose tuition is covered by an endowment or a special program to nearly $50,000 a year at some of the most exclusive college preparatory schools in the United States.

Another subclass of private schools consists of those that are supported by a particular religious group or denomination. These schools add instruction in religion to the curriculum and modify instruction in the regular content to highlight particular aspects of the religious group's or denomination's history or beliefs. These schools include parochial schools, which are established to educate the children of Roman Catholic families living in a given parish. Other religious bodies and denominations also sponsor private schools in addition to the Roman Catholic private schools.

A final form of schooling is the home school, in which parents assume direct responsibility for the education of their children, with very few strictures placed on them by the state in which the home school exists. In some cases, parents who are homeschooling their children have banded together to achieve the economy of scale and resources gained by having a critical mass of students.

In the 2013–14 school year, approximately 33,366 private schools were in operation, adding to the numbers of public schools and enrollment data shared above,. The organizing structure was Catholic for 21% of these schools, other religious bodies for 47%, and nonsectarian for 32% (Snyder and Dillow 2015).

The academic year 2011–12 is the latest year for which we have complete data on student enrollment in the entire educational enterprise in the United States, since census data lag school-year data by 3–4 years. In 2011–12, U.S. K–12 public schools accounted for more than 49,000,000 students. Private elementary and secondary schools contributed another 5,250,000 students, and homeschooling accounted for approximately 1,700,000 additional students (Snyder and Dillow 2015). Thus, these estimates give us slightly more than 55,950,000 students who were involved in K–12 educational programs in the United States in the 2011–12 school year. Projections for the 2015–16 school year suggest that the number of students involved in K–12 public and private education may have been around 55,957,000 students, composed of 50,773,000 public school students and 5,183,000 students in private schools (Hussar and Bailey 2013). This number does not include estimates of the homeschooled youth in K–12, who could add more than 2,000,000 students to the total for the 2015–16 school year (Ray 2015). (Note that because federal education data in the United State usually lag two to three years behind the date of their release, it is sometimes necessary in this report to speak of data from time that is past as estimates.)

At the postsecondary level, at the beginning of the 2011–12 academic year, 15,110,196 students were enrolled in public degree-granting institutions, and 5,883,917 were enrolled in private degree-granting institutions. Analyzing these data more closely, we find that the students enrolled in public institutions were divided, with 8,047,729 at four-year colleges or universities and another 7,062,467 at two-year colleges. A different distribution existed at private institutions, where the students were also divided, with 5,446,402 in four-year colleges and universities and another 437,515 studying in two-year colleges. Breaking down this latter number of students in private two-year colleges, we see that 29,864 of them were enrolled in nonprofit two-year programs, whereas 367,651 were enrolled at profit-making two-year institutions. The much larger number enrolled at these institutions is most likely a reflection of vocational programs offered at the two-year profit making institutions (Snyder and Dillow 2015).

As table 1 suggests, the number of students in public U.S. K–12 and postsecondary education has risen steadily since 1985 (Snyder and Dillow 2015). Projections through 2022 show the total number of students in K–12 public schools continuing to increase.

Table 1

School and postsecondary enrollments and projections over time (in millions)

	Year								
Type	**1985****	**1990****	**1995****	**2000****	**2005****	**2010****	**2015*****	**2020*****	**2022*****
K–12 public	39.4	41.2	44.8	47.2	49.1	49.5	50.3	52.1	53.0
K–12 private*	5.6	5.6	5.9	6.2	6.1	5.4	5.0	4.9	5.0
Postsecondary	12.2	13.8	14.3	15.3	17.5	21.0	21.3	22.8	23.5

*Nongovernmental, including parochial schools (governed by religious bodies).

**(Snyder and Dillow 2015).

***(Hussar and Bailey 2014).

Total public and private elementary and secondary school enrollment reached 55 million in 2005, representing a 22% increase since fall 1985. Between fall 2005 and fall 2015, a further increase of 1.4% was expected, indicating a slower rate of growth. When the data become available, increases in public school enrollment are expected in the proportions of Hispanics, Asians/Pacific Islanders, and American Indians/Alaska Natives, and decreases are expected to be found in the proportions of Whites and Blacks. Increases in public school enrollment are expected in the South and West, whereas decreases are expected in the Northeast and Midwest (Aud et al. 2011, Hussar and Bailey 2011, 2013).

Admission to Postsecondary Institutions

Graduates of public or private senior high schools may matriculate to the nation's colleges, but they must apply to the individual schools to be considered for admission. Most state-supported, two-year colleges will accept any secondary school graduate from the geographic area that they serve. Other two-year colleges and most four-year colleges require applicants for admission to have completed a specified number of courses in English, mathematics, science, social studies, and foreign language and to have a high school diploma. Many state-supported institutions have formulas for admission that may take into consideration the intended field of study, secondary school course grades, percentile rank in class, scores on college entrance examinations, letters of recommendation, participation in sports and other extracurricular activities, and other information supplied by a students' high school. Private colleges use some of the same criteria as public institutions but may consider factors such as whether members of the applicant's family have graduated from the institution. Very selective schools may consider the level of difficulty of courses taken in high school and the scores that applicants have earned on recognized Advanced Placement examinations, possibly using them to award acceptance with advanced standing on entrance. The mean costs of college undergraduate attendance, including tuition, fees, room, and food for in-state students at four-year public and private nonprofit colleges in 2005–6 (in current dollars) were $13,828 and $30,725, respectively (College Board 2015h; Glinder, Kelly-Reid, and Mann 2015). These totals increased by the 2010–11 year to $16,527 and $34,764, respectively. Adding another five years of growth to the costs, the four-year public and nonprofit private college costs for tuition, fees, room, and board in 2015–16 were $18,198 and $37,392, respectively.

Although many undergraduate students receive scholarships and other types of financial aid from various sources, including the college that they attend, government programs, or private foundations, the costs of attending a college and university are increasingly beyond the reach of many students and their families (College Board 2015h). The College Board estimates that full-time undergraduate students at private nonprofit schools received an average of about $18,870 in grant aid and federal tax benefits in 2014–15 to help pay their way through a year at their school. Full-time in-state students at public universities received an average $6,110 to assist in meeting their costs in the same academic year. The cost of attending two-year colleges varies widely, depending on the program selected by a student. In some cases, almost all expenses are borne by the local taxing district; in other cases, the costs are equivalent to those of a public four-year college or university.

The Every Student Succeeds Act of 2015

Because the Constitution of the United States does not claim education as a responsibility of the federal government, individual states have considerable leeway in structuring the education of their students. State laws define the boundaries for the compulsory education of students; outline the general framework for required studies in reading, writing, mathematics, science, social science, physical education, and other subjects; define the minimum number of days of school attendance per year; and define the standards for teacher certification and professional development. These laws, however, stipulate little or no regulation or monitoring for homeschooling. State laws also provide the mechanisms by which local schools are recognized by the state government and provide statutes for the founding and accreditation of private schools. In like manner, states have considerable leeway in waiving regulations for charter schools. These schools thus receive public funds but are not responsible for meeting all the regulations binding other public schools in the state or district.

The United States Department of Education sets standards and provides federal funding for special programs, such as school lunch programs for students in poverty and compensatory programs for students needing special educational assistance. The role of the federal government in education has increased markedly since the establishment of the No Child Left Behind Act (NCLB), passed by Congress in 2001. NCLB authorized the U.S. Department of Education to manage a program that provided financial incentives for schools with good performance profiles and penalties for schools with poor performance records. The program was unprecedented in the nation's history (U.S. Department of Education 2008).

Three days after taking office in January 2001, President George W. Bush announced No Child Left Behind, his framework for education reform that he described as "the cornerstone of my administration" (Bush 2009). Less than a year later, the United States Congress passed the No Child Left Behind Act of 2001. NCLB had four main thrusts: increased data-driven accountability for states, school districts, and schools; greater choice for parents and students, particularly those attending low-performing schools; more flexibility for states and local educational agencies (LEAs) in the use of federal education dollars; and a stronger emphasis on reading, especially for the youngest children. The disaggregation of state and local data required by NCLB mandated that all students, and in particular, special education students of various types, receive a high-quality mathematics education. In short, the law embodied the idea that the success of all students does truly mean a focus on *all*. In particular, the focus was on making every student proficient, according to the state equivalents of National Assessment of Educational Progress (NAEP) proficiency levels, described in Chapter 4.

The No Child Left Behind Act was set to expire on Sept. 30, 2007, inasmuch as the U.S. Congress passes laws with the intent that they will expire after a fixed period of time—most often, five years. This scheduled expiration is supposed to force Congress to update or amend a law, with some history of implementation to back up changes. However, if Congress somehow doesn't get around to taking a new look at a law, the law's authority doesn't go away; it remains intact until a new law is passed.

The No Child Left Behind Act had not been reauthorized prior to the beginning of Barack Obama's first term as president of the United States. In March 2010, the Obama administration released a blueprint for reform of the Elementary and Secondary Education Act, using the name by which the law has been formally known since its first passage in 1965, instead of No Child Left Behind, as the 2001 reauthorization under the Bush administration had been called (Duncan 2010). This blueprint recommended that states implement a broader range of assessments to evaluate advanced academic skills, develop and implement plans for the use of technology across the curriculum and in assessment, and foster students' capabilities to communicate effectively in writing and speaking. Other goals included engaging students in conducting research, using technology, engaging in scientific investigation, and solving problems effectively.

At the same time, President Obama proposed that the stringent accountability penalties, based on what NCLB called annual yearly progress (AYP), determined by the percentage of students at or above the proficient level in the NCLB legislation, be relaxed in favor of focusing more efforts on student improvement. Such improvement measures would involve modifying assessments appropriately for English language learners, minorities, and special needs students. In addition, school programs were to be revised to consider measures beyond reading and math tests. The blueprint suggested that the enactment of Obama's changes would retain students in school through graduation. Obama's plan also focused on closing the achievement gap between Black and White students.

Attempts to revise the bill foundered during Obama's first term as president and on into his second term but regained traction in the fall of 2014, and in the spring of 2015 separate bills were approved for reauthorization by the U.S. Senate and U.S. House of Representatives. The draft bills went into committee to be merged into a compromise bill that might pass both houses and be forwarded to the president. The resulting proposed bill, the Every Student Succeeds Act (ESSA), came out of committee on November 30, 2015, and was passed by both houses of Congress with strong bipartisan support. With President Obama's signature on December 10, 2015, the bill became the law of the land, replacing the No Child Left Behind Act of 2001, eight years after its scheduled reauthorization.

ESSA differs from NCLB in several significant ways. First and foremost, the new law shrinks the federal footprint, limiting the intervention of the U.S. Department of Education, and specifically the secretary of education, into most aspects of standard setting, assessments, and improvements based on standards. The bill retains the annual testing of students from grades 3 through 8 and at grade 11 in reading and mathematics. However, states will set their own standards and performance goals. They will still have to report the results of student performance assessments relative to state definitions of performance, provide data on student performance as measured by state achievement levels, and submit breakdowns in performance by gender, racial/ethnic cultural group, and disabilities, as required by federal law.

Furthermore, states will be freed of the pressures of the AYP strictures of NCLB but will still be required to provide interventions for the lowest 5% of school performers, for schools with high dropout rates, and for schools with persistent achievement gaps. The shift away from the AYP requirements removes the need for states to seek waivers to avoid federal penalties or interventions when they fail to meet federal performance standards. All states will now be charged with refocusing their efforts on helping students truly succeed in school rather than simply pass an examination. The law also separates accountability for test results from teacher merit evaluations—a key point that opponents fought to maintain. One other key point lost by the minority in the final form of Every Student Succeeds Act was the portability of Title I funding for students with disabilities. The minority argued that such funds should be tied to the students, to enable them to move to other schools of their choosing with the funds moving with them.

The prospect of maintaining a testing program without the federal strictures on performance and accountability to the federal government, as set out in the new act, causes some educators to decry the loss of a single, uniform federal standard for judging achievement. One educator has noted that students who do not come from a privileged family should recognize that they are on their own if they live in a state that does not believe in providing an excellent education.

The Every Student Succeeds Act puts targets in place for the appropriation of funds for the various parts of the law and promises to bring some stability to K–12 education policy until the bill comes up for reauthorization in 2020.

Chapter 2: The Intended Curriculum in an Age of Standards

The period from the release of the findings of the First International Mathematics Study (Husén 1967) to the release and continuing development of the Common Core State Standards for Mathematics (NGA Center and CCSSO 2010) has been the most sustained period of K–12 curricular focus in mathematics education in U.S. history. This recent period of intense interest in mathematics education emerged from earlier attempts to improve mathematics education in the United States, including the founding of the University of Illinois School Mathematics Program in 1951 (UICSM Staff 1957), the "New Math" era and the surrounding debates in the 1960s, and the "Back to the Basics" movement in the late 1970s.

By 1980, the leaders of the mathematics and science communities saw that a co-ordinated effort was necessary to ensure that the nation's youth were prepared for the demands of the world in which they would spend their lives. For that reason, this report takes 1980 as the starting point of the time period that is solidly tied to programs that are ongoing in mathematics education. In 1980, the National Council of Teachers of Mathematics released the report *An Agenda for Action: Recommendations for School Mathematics of the 80s*, which set forth eight recommendations for strengthening the school mathematics curriculum and the teaching of it (NCTM 1980). These recommendations were seconded by two ground-breaking reports released in 1983: *A Nation at Risk*, released by the National Commission on Excellence in Education (NCEE 1983), and *Educating Americans for the 21st Century*, released by the National Science Board's Commission on Precollege Education in Mathematics, Science and Technology (National Science Board 1983). These reports received front-page media attention and launched the move toward professional standards for the mathematics education community that culminated in 1989 in *Curriculum and Evaluation Standards for School Mathematics*, published by the National Council of Teachers of Mathematics (NCTM 1989). Thus, in effect, 1989 ended the first period that was devoted to shaping the vision that continues to inform mathematics education today.

The next period of growth and change spanned the period from 1990 to 2007. These years witnessed the broadening of NCTM's Standards for school mathematics to include standards for instruction, teacher education, and assessment. The NAEP mathematics assessment developed an extension of the NAEP reporting to include the issuance of state-by-state reports on students' mathematical achievement. This period saw the adoption, adaptation, and implementation of the NCTM Standards into the curricula of U.S. classrooms. However, this implementation, while coordinated within states, failed to achieve the desired coordination across states at grade levels.

In 2000, NCTM released a landmark publication, *Principles and Standards for School Mathematics*, which merged the Council's earlier curriculum standards efforts, with some elaborations and extensions. This event coincided with the release by the International Association for the Evaluation of Educational Achievement (IEA) of the results of the Third International Mathematics Study. The examination of the U.S. curriculum in this study was less than positive (Mullis et al. 1998). Afterward, commentators often referred to the U.S. curriculum as being "a mile wide and an inch deep" (Schmidt 2000). Thus, most of the first decade of the 2000s was committed to improving curricula and developing grade-level suggestions for content for portions of the curriculum, such as statistics. This period of reform ended in 2009.

The third time frame encloses the period from 2010 to the present, beginning with the release of the Common Core State Standards for Mathematics (CCSSM), embodying its vision of a "national" set of classroom standards and strengthened by its link with the governors and superintendents of public instruction in the states ratifying the CCSSM outcomes (NGA Center and CCSSO 2010a, 2010b). At present, 43 states and the District of Columbia have accepted CCSSM as their mathematics framework in part or in total. Extending on into the present, this period also encompasses the reauthorization of the No Child Left Behind Act of 2001 (NCLB) in the form of the Every Student Succeeds Act of 2015 (ESSA). The new law removes some of the punitive AYP measures that NCLB tied to the testing of student achievement, although it also loses the federal initiatives to effect change in mathematics outcomes at the state and local level. Overall, however, ESSA's innovations tend to support improved conditions and increased funding for mathematics education initiatives for students and their schools.

With this sketch of the three time periods in mind, let's look in detail at the evolution of the present standards for school mathematics.

1980 to 1989: Preparing for Change through Standards

From the release of the FIMS achievement results in 1967 indicating that U.S. K–12 students did not measure up in mathematics achievement with their peers in other countries, the American public began to take notice of U.S. students' mathematics preparation and consider the nation's competitiveness from an economic perspective (Hechinger 1967).

The release of NCTM's *An Agenda for Action* in April 1980 set the stage for a proactive era of professional input to the reform of mathematics education in the United States. This proactive period extends into the present. The Agenda outlined eight actions to consolidate the improvement of school mathematics in the decade of the 1980s. These recommendations took the form of tasks to be accomplished in reforming mathematics education across the 1980s. The report recommended that—

- *problem solving be the focus of school mathematics;*
- *basic skills in mathematics be defined to encompass more than computational facility;*
- *mathematics programs take full advantage of the power of calculators and computers at all grade levels;*
- *stringent standards of both effectiveness and efficiency be applied to the teaching of mathematics;*
- *the success of mathematics programs and student learning be elevated to a wider range of measures than conventional testing;*
- *more mathematics study be required for all students and a flexible curriculum with a greater range of options be designed to accommodate the diverse needs of the student population;*
- *mathematics teachers demand of themselves and their colleagues a high level of professionalism;*
- *public support for mathematics instruction be raised to a level commensurate*

with the importance of mathematical understanding to individuals and society (NCTM, 1980, p. 1).

These recommendations for action came at the end of a period of national dissatisfaction with the outcomes of schooling and a national divide over teaching basic facts versus problem solving and the use of technology in the classroom.

An Agenda for Action provided a framework for a structured discussion of the direction in which K–12 mathematics education should go in the coming years. The points were widely circulated and discussed, and they brought significant attention to the improvement of K–12 mathematics at state meetings of mathematics teachers and among faculty at colleges and universities working with teachers of mathematics.

In September 1982, the Conference Board of the Mathematical Sciences (CBMS) hosted a meeting of leaders of the mathematical science and mathematics education communities to discuss the topic "The Mathematical Sciences Curriculum K–12: What Is Still Fundamental and What Is Not." A CBMS report bearing the same name was released in December of that year. It outlined recommendations for elementary and middle school mathematics; traditional secondary school mathematics; non-traditional secondary school mathematics; the role of technology; relationships with other disciplines; and teacher supply, education, and reeducation. These boundaries spoke to what should remain and what should go from mathematics education at that time. They also spoke to what should be added, as well as where and how. But what was most significant at the meeting were the sidebar conversations focused on plans to extend the conversations of the CBMS meeting.

On April 26, 1983, before suggested changes could be implemented, the National Commission on Excellence in Education released its transformative report, *A Nation at Risk: The Imperative for Educational Reform* (NCEE 1983). The report began with the observation that the nation was at risk of losing its preeminence among the world's nations in commerce, industry, science, and technology, and it detailed the reasons for the erosion of U.S. leadership in these areas. In particular, it pointed to the decline of support for the basic purposes of schooling and for high expectations and the commitment needed to achieve them. Most famously, the report contended, "If an unfriendly foreign power had attempted to impose on America the mediocre educational performance that exists today, we might well have viewed it as an act of war" (NCEE 1983).

The report went on to describe the nature of the risk and its indicators, the steps that were necessary to recover excellence in education, and the actions that this journey would require of the public and those involved in education and its support. In particular, the report called for mathematics curricula in high schools to provide programs that—

> would equip graduates to: (a) understand geometric and algebraic concepts; (b) understand elementary probability and statistics; (c) apply mathematics in everyday situations; and (d) estimate, approximate, measure, and test the accuracy of their calculations. In addition to the traditional sequence of studies available for college-bound students, new, equally demanding mathematics curricula need to be developed for those who do not plan to continue their formal education immediately (NCEE 1983).

The recommendations called for increased entrance requirements to universities, achievement tests to document accomplishments at major transitions in the education process, increased time on task to allow students to meet the increased expectations established for them, enhanced expectations and rewards for teachers of mathematics and the sciences, special resources directed toward easing the shortage of mathematics and science teachers in the nation's schools, and assistance for school leaders and legislators responsible for funding and achieving these goals to help them understand their importance and ensure the fiscal and school environmental stability necessary to do so (NCEE 1983).

A look at the specifics of what needed to be done followed on September 12, 1983, in the report of the National Science Board's Commission on Precollege Education in Mathematics, Science, and Technology. The commission made its report to the National Science Foundation's National Science Board, and its title communicated its vision, scope, and sense of urgency: *Educating Americans for the 21st Century: A Plan of Action for Improving Mathematics, Science, and Technology Education for All American Elementary and Secondary Students So That Their Achievement Is the Best in the World by 1995.* It outlined plans for achieving its goal, and its mathematics recommendations fleshed out the recommendations made by the CBMS meeting of 1982 mentioned above.

It was clear now that change was afoot. Two additional meetings, both originating in the sidebar discussions at the 1982 CBMS meeting, were held late in 1983 and produced reports that were released in 1984. The major outcomes of the first one, which was supported by NSF, hosted by CBMS, and attended by leaders of the mathematical sciences and mathematical education communities in November 1983, were recommendations for national standards for K–12 mathematics and an ongoing Mathematical Sciences Education Board (MSEB), to be located at the National Research Council in the National Academy of Sciences. The second meeting spawned by the CBMS meeting in 1982 was held at the University of Wisconsin in December 1983, sponsored by the U.S. Department of Education's Office of Educational Research and Improvement (OERI). The participants represented school districts, state departments of education, national educational organizations, and, in one case, textbook publishers. At this second meeting, the calls for change focused on the development of standards, teacher education, needed research to back any standards, and a national steering committee like the proposed MSEB.

The reports of these two meetings, *New Goals for Mathematical Sciences Education* (*CBMS 1984*) and *School Mathematics: Options for the 1990s* (Romberg 1984), respectively, provide a clear picture of the tenor and recommendations emanating from the two conferences. Even more important, the meetings established a cadre of like-minded individuals who represented the mathematical sciences community broadly as well as various levels of the teaching of mathematics. Given the work that had taken place in committees of NCTM on the setting of standards, NCTM now felt empowered to move forward.

Various NCTM committees recommended that NCTM generate professional standards as a way of promoting excellence in the teaching of K–12 mathematics while discouraging unacceptable practices, much as the disciplines of engineering, medicine, and accounting had done. One area for proposed standards was the professional development of preservice teachers—an area in which the Council already had some experience through its work with the National Council for Accreditation of Teacher Education (NCATE), now known as the Council for the Accreditation of Educator Preparation (CAEP). Another area proposed for standards was the development of instructional materials designed for use in school mathematics classrooms. Proactive moves on the part of the Council to strengthen the teaching and learning of mathematics clearly had support from a number of directions.

The final decision by the NCTM Board of Directors came in March 1986 in the passage of a motion to begin "the development and implementation of professional standards for mathematics education in grades K–12." The first set of standards was limited to the content of the K–12 curriculum and statements about related changes necessary in the evaluation of learning environments and instructional materials. This decision to limit the scope of the first NCTM standards document reflected the belief that teachers, administrators, and the public could most easily relate to content and set to work on revising what they taught. Standards dealing with the professional development and teacher certification were put on hold, pending acceptance of the content recommendations. McLeod and colleagues (1996) provide a full discussion of the process of structuring the writing and vetting of the first set of standards.

The writing groups went to work in June of 1987, developing a draft and soliciting reactions from NCTM Board members and other leaders in the mathematical sciences and mathematics education community along the way. A central question at this stage was, What is a "standard"? The definition eventually developed was, "A standard is a statement that can be used to judge the quality of a mathematics curriculum" (NCTM 1989, p. 2). At the end of the summer of 1987 an interim draft, revised in response to reviewers' recommendations, was circulated at NCTM regional meetings and state mathematics education meetings and through special sessions devoted to securing feedback from teachers of the various grade levels during the 1987–88 school year. Presentations were also made to the annual meetings and boards of the major mathematical sciences professional organizations. All reactions were transmitted to the writers for consideration as they produced the final draft version that went to the NCTM Board in the fall of 1988. Following its approval as *Curriculum and Evaluation Standards for School Mathematics*, NCTM made plans for its release at a press conference in Washington, D.C., on March 21, 1989, some three years after the beginning of the project.

Considerable work went into ensuring that the release communicated the vision of school mathematics embodied in the Standards to a broad public, not as a rejection of content but rather as a reshaping of it to meet the needs of children in the emerging world. Two months before the release of the Standards, NCTM's initiative received a boost from the release of MSEB's *Everybody Counts* (1989), which highlighted the need for standards and provided a great deal of data building the case. The release of MSEB's publication *Reshaping School Mathematics* (1990) in the summer of the following year outlined steps that school districts could take in beginning to make change in their programs.

The reaction to NCTM's release of the 1989 Standards was very positive, and NCTM immediately began planning to extend the Standards efforts to teaching and assessment. The first step, though, was to set in place a nationwide program of professional development focused on the needed changes in state policy, curricular guides, and assessments that would support the recommendations in *Curriculum and Evaluation Standards for School Mathematics*.

NCTM mailed copies of the Standards to all NCTM members and distributed an Executive Summary to leaders of all the major mathematical sciences professional organizations as well as members of their boards. All members of Congress received copies, as did congressional staff members of all influential education- and science-related committees. The same mailing then blanketed department chairs at the university and secondary levels and influential members of the mathematics and teaching communities within states.

1990 to 2009: Adopting, Expanding, and Implementing the Standards

NCTM immediately began work on the development of *Professional Standards for Teaching Mathematics*, released in 1991, and *Assessment Standards for School Mathematics*, released in 1995. These two new Standards documents completed the picture of school mathematics while teachers were immersed in a period of adoption, refinement, professional development, and development of curricular materials in response to *Curriculum and Evaluation Standards for School Mathematics* (McLeod et al. 1996).

NCTM joined with allied organizations in mounting projects in support of the Standards. The Association for State Supervisors of Mathematics (ASSM) took the lead, launching a huge project, Leading Mathematics into the 21st Century, which also involved the National Council of Supervisors of Mathematics (NCSM), the Mathematical Sciences Education Board (MSEB), and the Council of Presidential Awardees in Mathematics (CPAM). Supported by funding from the National Science Foundation (NSF), the project organized and delivered five regional conferences to train professional development teams from states. It supplied them with direct access to the developers of NCTM's Standards, sample presentations that they could use in working with teachers in relation to the Standards, and a package of print and media materials for use as part of such presentations. These presentations focused on training state-based teams that included a business/industry leader, at least one school/state board member, K–12 teachers, a university mathematics educator and a university mathematician, and others.

To ensure equitable outreach to students and teachers, the numbers of state-based teams were proportional to the states' numbers of congressional districts. These teams returned to their states and trained other teams, with the goal of providing every teacher of mathematics, K–12, in every state with an opportunity to attend a program about the Standards in his or her own locality. These workshops were estimated to have reached more than 150,000 people in the first year. The activities also carried forward into the subsequent Standards releases with appropriate audiences and prepared the way for the release of *Principles and Standards for School Mathematics* in 2000. People accepted the recommendations from NCTM as trusted guides to school curriculum improvement in mathematics (McLeod et al. 1996). NCTM also initiated its largest-ever print-based professional development series, the *Curriculum and Evaluation Standards* Addenda Series, providing specific, grade-band examples of lessons and approaches to teaching topics aligned with the recommendations of the Council's first major Standards publication.

In this period, state standards underwent a significant shift as their curricular content recommendations moved toward alignment with those espoused by the NCTM Standards. But while the content taught at the same grades in schools within a state might exhibit little variance, what was taught in the same grades across states exhibited far greater variance. At the same time, strides made at the two-year/community college level brought more coherence to these institutions' curricula and focused more attention on their instruction, while tying the outcomes to career- and college-readiness outcomes beyond secondary school.

These efforts continued throughout the 1990s and into the middle of the next decade, with initiatives from related organizations playing major roles in providing information on how to meet the needs of students and chronicling the positive changes in student achievement. Central to these efforts were the following publications:

1995: American Mathematical Association of Two-Year Colleges (AMATYC), ***Crossroads in Mathematics: Standards for Introductory College Mathematics***

TIMSS: *Third International Mathematics and Science Study* (follow-up reports in Beaton et al. [1997]; Mullis et al. [1997, 1998, 2000, 2004, 2008a, 2012])

2000: National Commission on Mathematics and Science Teaching for the 21st Century (Glenn Commission), *Before It's Too Late*

NCTM, *Principles and Standards for School Mathematics*

2001: CBMS, *The Mathematical Education of Teachers*

National Research Council, *Adding It Up: Helping Children Learn Mathematics* (Kilpatrick, Swafford, and Findell)

2004: Organisation for Economic Cooperation and Development (OECD), *Learning for Tomorrow's World: First Results from PISA 2003* (follow-up reports in OECD PISA [2004a, 2007, 2010, 2013a, 2013b, 2014])

OECD, *Problem Solving for Tomorrow's World: First Measures of Cross-Curricular Competencies from PISA 2003* (OECD PISA 2004b)

MAA Committee on the Undergraduate Program in Mathematics (CUPM), *Undergraduate Programs and Courses in the Mathematical Sciences* (MAA 2004a, 2004b)

2005: American Statistical Association (ASA), *Guidelines for Assessment and Instruction in Statistics Education: College Report* (Aliaga et al. 2007; this publication provided a vision of an introductory statistics course and the instruction and assessment that might embody that vision)

2006: AMATYC, *Beyond Crossroads: Implementing Mathematics Standards in the First Two Years of College*

College Board, *College Board Standards for College Success: Mathematics and Statistics*

NCTM, *Curriculum Focal Points for Prekindergarten through Grade 8 Mathematics*

2007: American Statistical Association (ASA), *Guidelines for Assessment and Instruction in Statistics Education: A Pre-K–12 Curriculum Framework* (*GAISE Report*) (Franklin et al. 2007)

NCTM, *Mathematics Teaching Today: Improving Practice, Improving Student Learning*

2008: National Mathematics Advisory Panel, *Foundations for Success: The Final Report of the National Mathematics Advisory Panel*

2009: NCTM, *Focus in High School Mathematics: Reasoning and Sense Making*

This list illustrates the massive amount of work focusing on the improvement of prekindergarten to collegiate teaching of mathematics during this period. A perusal of the list indicates that the idea of standard setting was moving into the community college and collegiate curricular discussion. The appearance of the CBMS's volume on the mathematical education of teachers called for a balance of preparation in content and in instruction and provided guidance in constructing programs to help preservice and in-service teachers learn in the ways that they themselves will need to teach.

It is evident that the growth of international comparative studies of mathematics and science will be a continuing force in discussions of curricular sufficiency. These studies include, but are not limited to, the quadrennial Trends in International Mathematics and Science Study (TIMSS), launched by the International Association for the Evaluation of Educational Achievement (IEA), and the triennial Programme in International Student Assessment (PISA), launched by the Organisation for Economic Cooperation and Development (OECD). These two assessment programs provide a platform for watching the flow of students through a variety of curricula worldwide and supply a great deal of contextual information about the programs and cultural factors in each country. They serve as steady partners to the U.S. program, the National Assessment of Educational Progress (NAEP), administered by the U.S. National Center for Educational Statistics (NCES), with data going back to 1973. Chapter 4 examines data from all these assessments in viewing the attainment of students passing through U.S. educational programs.

NCTM's *Curriculum and Evaluation Standards for School Mathematics* (1989), followed in the early 1990s by its partners that articulated teaching and assessment standards, gave focus to content standards across the three grade bands K–4, 5–8, and 9–12 for the processes of problem solving, communication, reasoning, and connections. NCTM restated the 1989 Standards in 2000 to reflect the growth of knowledge about learning and practice over the intervening period of time. In addition, this revision and updating, published as *Principles and Standards for School Mathematics* (NCTM 2000), sharpened the original 1989 recommendations by unifying content, its teaching, and assessment into a single picture of three mutually supportive domains that make a mathematics program whole. The revision offered the following major refinements :

- Shifted the suggested content placements from three grade bands, K–4, 5–8, and 9–12, to four grade bands: pre-K–2, 3–5, 6–8, and 9–12
- Added the process of representation to the group of mathematical processes addressed in the Process Standards
- Made specific suggestions for content for the grade bands addressed
- Merged the previous NCTM Standards on teacher education, professional development, assessment, and evaluation with curricular recommendations into a single document.

The release of *Principles and Standards* reignited the movement to pull state standards for mathematics closer to the curriculum content espoused by the NCTM Standards. But, as mentioned earlier, although there might have been little variance in the content being taught at the same grades in schools within a state, there was far greater variance in what was being taught in the same grade levels between states.

This gap did not close. At the same time, two-year and community colleges were making strides to bring coherence to their curricula and reflect on the way in which instruction was taking place in their classrooms while tying the outcomes to applications in careers and paths to mathematical settings beyond secondary school.

These markers of college and career readiness were evident in the College Board's 2006 release of *Standards for College Success: Mathematics and Statistics*. This document outlined two curricular paths for students starting at the sixth-grade level and taking mathematics each year through high school graduation. On one path, students completed algebra 1 by the end of grade 8 and proceeded on a course track taking them to the equivalent of at least Advanced Placement Calculus or Advanced Placement Statistics by the end of grade 12. On the other path, students followed a curriculum that enabled them to complete three courses in middle school mathematics with a strong algebra core by the end of grade 8 and then take courses in high school leading them to the equivalent of precalculus in grade 12. The documents from the College Board Program also provided a listing of topics arranged by grade level and offering guidance for establishing these programs of study, either by a sequence of traditional courses or by a series of courses integrating the teaching of algebra, geometry, functions, statistics, and modeling over each of the four years of high school mathematics study (College Board 2006). A notable feature of these programs of study is grade-level expectations, rather than the grade bands of the NCTM *Principles and Standards*.

In that same year, NCTM published its *Curriculum Focal Points for Prekindergarten through Grade 8 Mathematics: A Quest for Coherence* (2006), which provides grade-level recommendations for the placement of topics and key concepts within the teaching of the topics at those grade levels. Although some believed that the focal points verged on the development of a national curriculum model, which NCTM had avoided in 1989, this NCTM initiative can be viewed as a response to the Council's recognition of the need to provide curricular coherence and clarity for students. The Focal Points were rooted in the same thinking that the College Board and the American Statistical Association were using in preparing standards documents for their special needs: a basis for assessment and the professional development of teachers and statistics education, respectively. *Curriculum Focal Points* focuses on coherence in the study of mathematics, created by a careful building of key concepts over time, interlinked so that each one builds on and extends the previous one, which paved the way to it. This linking of topics over grades provides the structural fiber that supports understanding mathematics as a discipline.

NCTM's release of *Curriculum Focal Points* was followed in 2007 by the American Statistical Association's (ASA) release of *Guidelines for Assessment and Instruction in Statistics Education* (*GAISE Report*): *A Pre-K–12 Curriculum Framework*. This document, like the former two, the College Board's *Standards for College Success* and NCTM's *Curriculum Focal Points*, provided developmental-level suggestions and activities for the teaching and learning of core statistical concepts and their applications.

The appearance in 2009 of NCTM's *Focus in High School Mathematics: Reasoning and Sense Making* rounded out NCTM's "focus" initiative with a publication examining a path for secondary school mathematics in the 21st century. Instead of focusing on content directly, as *Curriculum Focal Points* had done, this publication sought to bring focus to high school mathematics by preparing students for using their mathematical understanding in reasoning about and making sense of mathematical situations. Building on the earlier Principles and Standards enunciated by NCTM, this publication addressed the components of reasoning and illustrated how they apply in each of the content domains addressed in *Principles and Standards*. The examples in *Focus in High School Mathematics* contain vignettes that model the roles that students and teachers play in working with situations that call for mathematical sense making and reasoning to solve problems such as those that students could be expected to encounter in life. The publication also addresses issues that administrators, supervisors, and curriculum specialists will face in achieving the goals set when sense making and reasoning are given central roles in the secondary curriculum.

All the foregoing activities were generated from within the mathematical science community, in which mathematics education is a major player. The development of the NCTM Standards had been accomplished within this community, working with mathematicians, mathematics educators and teachers, administrators, and individuals from other disciplines.

In 2007 the structure of the development of standards got an unprecedented boost from individuals in state government. At this time, the National Governors Association (NGA) and the Council of Chief State School Officers (CCSSO) turned their attention to education at their respective annual meetings. Their focus was on bringing individual state standards into closer agreement with one another and building on the base of commonality at grade levels established by the NCTM Standards, other educational recommendations, and what was taught in many other countries that were worldwide leaders in academic achievement.

By this time, all the U.S. states had some form of state standards, but these exhibited significant inconsistencies in the grade levels at which specific topics were introduced and in the time allotted for students to develop fluency with procedural skills. In some cases, these levels were never specifically addressed, leaving the decisions to local districts or even schools (Reys 2006). These conditions moved the NGA Center for Best Practices and CCSSO to launch the Common Core State Standards Initiative in 2008 with the appointment of a writing group, a series of drafts and revisions, and the development of publications outlining the Common Core for state programs.

2010 and Beyond: CCSSM and Its Impact

In 2010, the National Governors Association Center for Best Practices and the Council of Chief State School Officers took the standards movement a step closer to a national curriculum with the release of the Common Core State Standards for Mathematics (CCSSM; NGA Center and CCSSO 2010a, 2010b). CCSSM provided a path for unifying the existing state standards and grade-level expectations for students into a suggested common set of goals for all states. The following sections of this chapter provide a bit more detail about the important documents that were developed along the way and an overview of the politics that determined a common set of expectations in school mathematics for all students. This effort built on the foregoing works defining the intended curriculum, several major reports, improvements in state assessments, and a growing body of research studies and curricular development and evaluation programs.

The Common Core State Standards for Mathematics (CCSSM)

CCSSM resulted from a state-led effort, initiated by state leaders, including governors and state commissioners of education. The Council of Chief State School Officers and the Center for Best Practices of the National Governors Association led the development of a K–12 curricular framework, drawn from the best of state standards, international curricular frameworks, and research results concerning mathematics teaching and learning, as well as teachers' experiences. The writing team consisted of mathematics content experts; mathematics educators and supervisors; assessment staff members from ACT, the College Board, and Achieve (an organization supporting standards-based education reform efforts across the states); and experienced teachers.

In December of 2008, NGA, CCSSO, and Achieve released *Benchmarking for Success: Ensuring U.S. Students Receive a World-Class Education*, which provided a basis for using international experience in curriculum construction as a model for bringing the state standards together. The Common Core State Standards for Mathematics elaborate grade-by-grade standards and related expectations for K–grade 8, as well as high school standards organized by conceptual categories that define the mathematics that students need to know to be college or career ready. The secondary school recommendations provide guidance for curricula delivered by the traditional topical design as well as curricula delivered in an integrated format (NGA Center and CCSSO 2010b).

The CCSSM Standards for Mathematical Practice complement the content domains and standards for K–8 and the subject-area conceptual categories for high school by describing processes and practices in which students should engage in "doing" mathematics and the expectations and outcomes that programs should hold for students. These Standards for Mathematical Practice are based on the NCTM Process Standards (NCTM 2000) and the levels of mathematical proficiency described in the National Research Council's *Adding It Up* (Kilpatrick, Swafford, and Findell 2001). The Standards for Mathematical Practice are as follows:

- Make sense of problems and persevere in solving them.
- Reason abstractly and quantitatively.
- Construct viable arguments and critique the reasoning of others.
- Model with mathematics.
- Use appropriate tools strategically.
- Attend to precision.
- Look for and make use of structure.
- Look for and express regularity in repeated reasoning.

Like the NCTM Process Standards on which they are based, the Standards for Mathematical Practice address the intellectual habits that a program needs to model for students to inculcate in them the associated behavioral and cognitive capabilities. CCSSM connects the mathematical practices with the structural goals of *coherence* and *focus*. *Coherence* refers to the vertical alignment of the content that prepares students for what is coming and assists them in reflecting back on the relationship between what they are currently studying and what they have previously studied. The removal of some topics from the traditional content across the K–12 curriculum opens space and time for students to focus on fewer topics in greater depth, developing their sense-making and reasoning capabilities as they work with the new material.

CCSSM was intended to move the nation to a common understanding of what children need to know and when, even if the standards were not identically the same from state to state. That is, the goal was to increase the ability of the system to deliver high school graduates who were college and career ready in mathematics, according to an explicit description of what this means. "Raise the level of rigor" was a phrase often used to indicate the intent of CCSSM, but it was an inaccurate description of the goal, which was to ensure that students understand mathematics as structure, reasoning, and procedures that make sense (rather than as a long list of rules to follow) and that they acquire the ability to reason with and use mathematics. As such, CCSSM was envisioned as a set of standards that would be adopted or adapted by the states, allowing all students to have equal access to similar instructional activities based on materials created to reflect a similar vision of school mathematics.

To date, keeping track of states signing on as adopters of CCSSM has been an interesting task. Initially, 45 states, the District of Columbia, and some of the U.S. territories adopted CCSSM as their official curriculum guide for mathematics. The states of Alaska, Minnesota, Nebraska, Texas, and Virginia initially chose not to participate in CCSSM, although Minnesota adopted the English Language Arts portion of the Common Core

State Standards. By the end of 2010, the CCSSM recommendations were adopted by 45 states and the District of Columbia. With the withdrawal of two of these states, as discussed later, the present total in 2016 stands at 43 states and the District of Columbia (Academic Benchmarks 2015).

Dissemination, Professional Development, and Opposition to CCSSM.

In July 2009, a year before CCSSM's release in June 2010, the U.S. Department of Education announced a competitive program, Race to the Top, which would award grants to states and consortia to adopt "internationally benchmarked standards and assessments that prepare students for success in college and the workplace." Although the criterion did not explicitly name CCSSM or require its adoption, grant applications from adopting states received extra points in the grant approval process if they had adopted CCSSM by August 2, 2010. This action on the part of the Department of Education, and hence the U.S. government, was viewed in many quarters as a highjacking of the Common Core Standards movement initiated by the states and without federal funding. This perception was strengthened by the funding of two state-based consortia for the development of assessments of Common Core curricula at state levels.

The assessment consortia are commonly referred to as PARCC (Partnership for Assessment of Readiness for College and Careers 2015a) and SBAC (Smarter Balanced Assessment Consortium 2015). The assessments had their full-scale field-testing in 2014–15 and will be serving as yearly statewide assessments in several of the participating states in the 2015–16 school year. In light of the lack of curricular materials developed and written explicitly to conform with the CCSSM framework, there is currently a disconnect between what is happening in implementing CCSSM and the mandated state assessments that still remain in the extant NCLB legislation.

CCSSM implementation challenges have come into the political spotlight. Some of the difficulties include the lack of curricular materials, the absence of widespread professional development for teachers, parental apprehension about what might be happening to their children's mathematical education, and the inability of the press and the public to separate the CCSSM initiative from the Department of Education's Race to the Top program. Already, political heat and the fear of a federal takeover of states' rights to determine their own educational structures have led the legislatures in Indiana, Oklahoma, and South Carolina to vote to withdraw from the list of early Common Core adopters. Other states have adopted "opt out" legislation, allowing parents to permit their children to be excused from having to take the tests. This leaves, at present, 43 states and the District of Columbia as adopters of CCSSM. Twenty-two of these states and the District of Columbia adopted the Common Core standards verbatim, while the remaining 21 states adopted them with minor modifications. Many of the non-adopting states have strong goals that parallel both the NCTM and CCSSM standards, and they have strengthened them over the years. These states were unwilling to move their teachers and schools through another curricular change for little perceived gain in alignment of their curricula or professional development programs.

Adoption of CCSSM has also been impeded by many forces presently acting on the nation's education system, including issues related to teacher retention; large numbers of beginning teachers in classrooms, often with the most challenging groups of students; political attacks on teachers; the perceived low quality of preservice teacher education programs; and an undervaluing of the profession of teaching—a climate giving rise to the question, What does setting certification requirements aside say about the standards states already have? All of these concerns created unease about what was happening in schools. The low esteem in which many Americans hold mathematics added fuel to the public angst about the adoption of CCSSM in many communities.

An important point deserves emphasis: Much of this angst was associated with the assessments, which were *not* a part of CCSSM! Much of the opposition to CCSSM is based on a misinterpretation of the standards and a misrepresentation of them as a federal government intrusion into the classroom. Much of both the misinterpretation and the misrepresentation stemmed from conservative groups using CCSSM as a tool in creating political rhetoric in election campaigns. A large amount of inaccurate and inflammatory information about CCSSM was spread by social media and was also linked to changes in assessment programs that expect students to explain answers to open-ended questions rather than just select or write down an answer.

Built around a core of mathematical practices and accompanied by specific mathematics content, CCSSM focuses on developing deep student understanding of a set of outcomes that is smaller than those currently contained in most state standards. This reduced number of outcomes in CCSSM reflects the fact that under previous sets of curricular recommendations, many standards were repeated in subsequent grades as the curricula on which these standards were built circled back to the same topics year after year. CCSSM tends to focus on core, or focal, topics for longer periods of study in a single year and then integrates further work with those topics with other disciplines.

Such linkages show the topics' importance and assist in anchoring students' learning more firmly in modeling and the mathematical practices. Chapter 6 illustrates and explains other aspects of the CCSSM recommendations.

Other mathematics education activities in the period starting in 2010 resulted from several other influential reports that appeared after the release of CCSSM. Several of these publications dealt with the integration of mathematics, science, and technology courses in preparing teachers, under the general designation of STEM-related activities, with STEM as an acronym for "science, technology, engineering, and mathematics." The uniting of these course areas has reflected a recognition of the need for a unified, integrated curriculum. Another part of this drive has resulted from the shortage of teachers to staff courses in mathematics and science classrooms in grades 7–12 across the country. The following two major reports detail this problem and methods of meeting it:

- *Successful K–12 STEM Education: Identifying Effective Approaches in Science, Technology, Engineering, and Mathematics* (NRC 2011)
- *Monitoring Progress toward Successful K–12 STEM Education: A Nation Advancing?* (NRC 2013b).

Along with these reports are two interrelated publications that deal with the changes taking place in teacher education in mathematics. The first publication deals with needed changes in the mathematics education of teachers, bringing the 2001 CBMS report on the topic up to date and calling on the entire mathematics community to get behind the continuing need for education for teachers of mathematics, K–12. The second publication reports on changes being made in the accreditation of teacher education programs for teachers of mathematics. These documents are as follows:

- *The Mathematical Education of Teachers II* (CBMS 2012)
- *NCTM 2012 CAEP Standards* (Council for the Accreditation of Educator Preparation; NCTM 2013)

The Mathematical Education of Teachers II is a reference for all who teach mathematics and statistics to preservice and in-service teachers. It calls for a solid education in mathematics for all teachers—an education that is linked to what preservice and in-service teachers will be teaching, not just a list of courses to check off as having done.

Further, the courses that the MET II report recommends for preservice and in-service teachers must provide learning situations that exhibit the mathematical practices, focus, and coherence that are called for in CCSSM. If it is true that one teaches as one has learned, then, the new standards assumed, the collegiate teaching of mathematics and in-service programs must change as well. MET II provides vignettes and other examples of how both content and mathematics-oriented pedagogy can be built into a program for educating all those who will teach mathematics.

The CAEP Standards were formerly known as the NCATE Standards, the standards that states and districts turned to in certification and hiring decisions related to teachers. NCTM has worked with CAEP over the years, and CAEP has adopted standards for recognizing teacher education programs developed by NCTM committees for some time. The 2012 set of CAPE Standards is just the most recent iteration of those standards, developed by a partnership that sees its work as extending beyond simply the stating the standards to ensuring that each visitation team has a committee member who is knowledgeable about the standards and what such a program looks like in action, not just on paper.

In 2014, NCTM published *Principles to Actions: Ensuring Mathematical Success for All*. This publication focuses on what success has looked like in mathematics education in the last decade, before addressing the inequities in how it has been dealt out to the nations' students by wealth, racial/ethnic/cultural origins, by community type, and by a myriad of other factors. The hope is that this time of curricular change may offer a chance for change in how instruction and opportunity are made available to all students, not just those who were born white and reside in a privileged neighborhood. The text elucidates the principles that we must identify and adhere to if we are to eradicate the barriers and ensure that all students have valid opportunities to learn mathematics. In particular, we must ask ourselves, do all students have access to knowledgeable teachers who are effective in using instructional modes that accommodate students who learn in different ways? Can we give assurances that the barriers that have in the past impeded equitable access to help, information, and the tools of learning in a modern mathematical setting have been dismantled? Is the curriculum, together with the supporting materials, equally available across all schools in all locations? If the Common Core State Standards are intended to ensure that all students will emerge from secondary education ready for college or career, do we have mechanisms in place to ensure that all will receive the equal opportunities that the standards promise? Are assessments developed in a way to be free of color bias and biases based on outside experience—that is, are they free of factors that lead to differential item functioning, depending on who a student is and where he or she comes from? Are students and teachers assigned to classes in a fashion that ensures that, as much as possible, competence and experience are shared equally across all students? Has the language in our schools changed from students' "abilities"—terminology that assumes that students arrive already ranked—to students' "capabilities"—terminology that focuses on and emphasizes what students can do and where programs can take them?

The years 2014 and 2015 also brought with them a focus on the educational programs that departments in mathematical sciences should offer for their students. Four publications are noteworthy in this regard:

- *Curriculum Guidelines for Undergraduate Programs in Statistical Science* (ASA 2014)
- *The Statistical Education of Teachers* (Franklin et al. 2015)
- *MAA-CUPM Curriculum Guide to Majors in the Mathematical Sciences* (MAA 2015a)
- *Guidelines for Assessment and Instruction in Mathematics Modeling Education* (SIAM and COMAP 2016)

In 2014, the American Statistical Association in 2014 released *Curriculum Guidelines for Undergraduate Programs in Statistical Science*, which outlines coursework and contents of courses in the statistical sciences. Accompanying white papers speak to various features of the guidelines and the teaching of statistics at the undergraduate level.

The following year brought ASA's release of *The Statistical Education of Teachers,* which parallels the coverage given in MET II while speaking to the well-developed philosophy of instruction in statistics developed by the statistical community over the past two decades. The Statistical Education of Teachers is covered in greater depth in Chapter 7.

MAA's recently published *CUPM Curriculum Guide to Majors in the Mathematical Sciences* is discussed at a later point in this chapter.

Guidelines for Assessment and Instruction in Mathematics Modeling Education is a new publication developed by the Society of Industrial and Applied Mathematics and the Consortium for Mathematics and Its Applications. It is still in development at the time of this writing but will be available before ICME-13 in Hamburg. Given the interest evinced by schools whose teams participate in poster competitions, developing and conducting statistical projects, and competing in the HiMAP competition each year, as well as other activities that involve mathematical modeling at the undergraduate level, the need for these guidelines is evident. Further, the individuals involved in developing them are very knowledgeable about such modeling activities, the school curricula, the CCSSM recommendations, and structuring modeling activities that are appropriate for students at various levels.

Changes in States' Secondary School Graduation Requirements in the CCSSM Era

Along with the adoption of CCSSM and the call for preparation for college and a new workplace for all students—preparation that requires more mathematics for all—states have been forced to reexamine their legislated high school graduation requirements. As of 2015, 47 states had statewide credit requirements for high school graduation, and the other 3 states were still reviewing their stance. At present these requirements, as determined by an exhaustive search of state documents in late 2015, can be categorized as follows:

- Thirteen states, plus the District of Columbia, have statewide credit requirements compelling students to earn four credits (years) in mathematics (consisting of algebra 1, geometry, and algebra 2, plus another advanced level course).
- Three states require algebra 2, geometry, and two additional mathematics courses.
- Ten states require students to complete algebra 1, geometry, and the equivalent of algebra 2.
- Nine states require students to complete three courses but stipulate only algebra 1 and geometry specifically by name.
- Four states require the completion of two courses—algebra 1 and geometry.

- Four additional states require two courses but, at most, named only algebra 1 as one of the required courses.
- Three states require passing a state-specified proficiency test in mathematics, usually resting on content equivalent to algebra 1, geometry, and some algebra 2 content.

Several states also stipulated that one of the required mathematics courses be taken in the final year of high school. The rationale for this requirement is tied to lessening the time gaps between secondary and tertiary experiences with mathematics.

State Standards and Textbook Adoption Processes

As a result of the release of the Common Core State Standards for Mathematics, official state standards are currently ahead of materials based on them for classroom instruction. Most of the major education publishing houses and media production companies in the United States are just releasing such materials or will be doing so in the next school year. Fortunately, the former state standards—based on *Curriculum Focal Points for Prekindergarten through Grade 8 Mathematics* (NCTM 2006), NCTM's secondary recommendations, and previous professional development based on NCTM recommendations—have enough overlap to help states and school districts begin the changes as they await new materials and professional development on newer portions of the CCSSM recommendations.

In essence, the change from the NCTM Standards–based curriculum to the CCSSM-based curriculum is more than an evolution in the U.S. school mathematics curriculum. The impact of the quick change has yet to be determined, since most teachers and schools have not had a chance to work through the full implications of the necessary changes to their curricula and assessment methods.

Historically, in most states, local school districts have made the end decisions regarding which instructional materials their classrooms would use. For textbooks, this decision process is called textbook adoption and may be subject to formal regulations. In 21 states, a portion of state education funds is earmarked for textbooks that are selected or recommended by statewide committees for use in that state's classrooms in accordance with the state's content standards. In these states, all adoptions for a given course or level may take place in the same school year. The State Instructional Materials Review Association (SIMRA) is a group of state officials charged with the adoption of textbooks in their state. Information about their activities can be found at SIMRA's website: http://simra.us. A 2011 communiqué from the state textbook adoption group in the California Department of Education to the board of the National Association of State Textbook Administrators (NASTA) indicates the pressure created by the release of CCSSM:

> California … is committed to implementing the Common Core State Standards (CCSS) adopted by the California State Board of Education (SBE) in August 2010. While it will take a number of years to develop new curriculum frameworks and instructional materials aligned to the CCSS, State Superintendent of Public Instruction Tom Torlakson has invited publishers of state-adopted programs in mathematics and language arts to submit supplemental instructional materials that bridge the gap between their existing programs and the CCSS. Select teachers and content experts will review the supplemental materials, and the California Department of Education (CDE) will list on its Web site those materials found to meet the evaluation criteria. The process is completely voluntary for publishers, and the materials will not be adopted by the SBE. Yet, this information should help school districts significantly in their transition to the CCSS in light of the

fact that the California State Legislature and Governor recently extended the suspension of SBE instructional materials adoptions until July 2015. (California Department of Education 2011)

The communiqué not only expresses the angst but also suggests the timeline disconnect that states face in adopting materials in real time. There is a real-time interplay between the development of materials and the preparation of the educational workforce to use them effectively as intended in the classroom. Reports from other large states in SIMRA on their current textbook adoption processes indicated similar interrupted circumstances in both states with state adoptions and those where local districts can choose their own materials independently.

Historically, the direction taken by the large textbook adoption states—California, Florida, and Texas—has had a large influence on the education materials market and choices for other states. In 2011, a Wisconsin foundation, the Brookhill Institute for Mathematics, joined with Texas Instruments, the Council of Chief State School Officers (CCSSO), and the National Council of Supervisors of Mathematics (NCSM) to develop a series of analysis tools for the adoption of mathematics instructional materials.

Hence, schools and school systems across the United States are slowly making significant headway toward the implementation of CCSSM. They do not want to make changes in their programs until decisions about their own state standards have been made by their state departments of education, which, with other states' departments of education, set the CCSS Initiative in motion in the first place (Association of American Publishers 2015).

With so many states signing on to CCSSM, what the state departments are going through at present what might be likened to a "Standards Spring," but it is not clear that they all have the same idea of what they are going to do with the new standards. The 43 states and the District of Columbia that have adopted CCSSM will now wield influence over textbooks and their approaches to the content, whereas in the past years only Texas, California, and a few other big states held sway. As noted earlier, some states have adopted the CCSSM verbatim, while others have adapted them, but not all in the same way. With the Every Student Succeeds Act superseding the NCLB legislation, will the states move to positions of greater uniformity in outcomes across states, or will some states revert to where they were before? Will the CCSSM program actualize its potential to become the national curriculum that it was envisioned to be? Will curricular materials be designed to support a more focused curriculum serving both career and higher education needs? Will the CCSSM movement also bring educational equity to the mathematics classroom? The answers to these questions reside in the decisions that state departments of public instruction make between this writing and the fall of 2016. In any case, careful plans for implementation will still need to be developed to bring the new curricula, instructional materials, and professional development to the nation's teachers, schoolrooms, and students.

College-Level Curricular Documents

The past decade has seen the focus in curricular documents related to collegiate mathematics shift from a listing of course contents to documents looking introspectively at content and appropriate instructional strategies for specific content in collegiate classrooms, as well as at ways in which the collegiate mathematics curriculum might be linked to the applications of mathematics in new disciplines.

Echoing some of the same themes as *Beyond Crossroads: Implementing Standards in the First Two Years of College*, published by the American Mathematical Association of Two-Year Colleges in 2006, the MAA Committee on the Undergraduate Program in Mathematics (CUPM) recently developed and published its own curriculum guide. The *2015 CUPM Curriculum Guide to Majors in the Mathematical Sciences* clearly expresses the systemic links among cognitive processes, mathematical practices, mathematics content, and the development of mathematical ways of knowing and applying the mathematical sciences in areas beyond the classroom. It also emphasizes the importance of developing model syllabi for courses interactively with colleagues within and outside one's own department or campus and elaborating how they fit together to provide programs of study matching students' goals for the future. The CUPM work extends the role of curriculum as it plays out in professional development, assessment, technology, undergraduate research, and a number of other facets of a successful program. Focusing on the interactive nature of curriculum in making choices, the CUPM work shows the diversity of the mathematical programs to which the committee's recommendations apply.

The CUPM report refers to two special publications of the National Research Council that provide assistance to a department in the development of undergraduate programs of instruction in mathematics. *Fueling Innovation and Discovery: The Mathematical Sciences in the 21st Century* (NRC 2012) and *The Mathematical Sciences in 2025* (NRC 2013a) both point to the roles that mathematics plays in the modern world. They focus on issues that departments of mathematical sciences should be thinking about as they seek to build capabilities within their faculty and to alert students to current and future applications and opportunities.

Alerting students to opportunities is key to recruiting and retaining them in STEM subjects as they move through our classes and programs. The President's Council of Advisors on Science and Technology (PCAST 2012) presented recommendations in its report, *Engage to Excel: Producing One Million Additional College Graduates with Degrees in Science, Technology, Engineering, and Mathematics*, with an eye to assisting secondary schools, colleges, and universities in recruiting and retaining capable students in fields in these areas. Such actions require shifting instructional programs to use validated instructional techniques and having the financial resources to make significant changes in classroom environments. These changes might include moving from lecture formats to settings that invite students to engage in group work, moving from inspecting tables and graphs in a book to using technology to represent and analyze data and interpret results, and considering community problems from a mathematical modeling standpoint as students prepare for careers in STEM fields or in mathematics and science teaching. Such changes also require concomitant changes in the instructional patterns in postsecondary institutions, especially at the undergraduate level, before these students are lost to other majors.

Supporting the ideas of PCAST in *Engage to Excel*, Project INGenIOuS (Investing in the Next Generation through Innovative and Outstanding Strategies) focused on development of the workforce. This endeavor was a joint project of the Mathematical Association of America, the American Statistical Association, the American Mathematical Society, and the Society for Industrial and Applied Mathematics. Many of the recommendations that emerged from this project have direct relevance to undergraduate instruction through faculty development and recruitment to STEM activities and majors. The project results can be found in the *INGenIOuS Project: Report on July 2013 Workshop* (MAA 2014). In summary, the workshop report called for six threads of activities, intertwined with one another:

- Bridge gaps between business, industry, and government (BIG) and academia.
- Improve students' preparation for nonacademic careers.
- Increase public awareness of the role of mathematics and statistics in both STEM and non-STEM careers.
- Diversify incentives, rewards, and methods of recognition in academia.
- Develop alternative curricular pathways.
- Build and sustain professional communities.

Although these recommendations are not new, they echo the recent call for new applications and needs in the post-baccalaureate world for a new type of mathematically trained and applications- and modeling-ready graduate. They address opportunities and challenges in mathematics that are well illustrated in a Society for Industrial and Applied Mathematics report, *SIAM Report on Mathematics in Industry* (2012), and in the American Statistical Association's *Teaching Statistics: Resources for Undergrad Instructors—More Data, Less Lecturing* (Moore 2001). These documents amply illustrate contemporary ways to take first steps to bring the wide range of applications and careers in the mathematical sciences to students.

A third effort, from the statistical sciences community, sought to bridge the disciplines and offer goals and methods for instructional change in an introductory statistics course. This work was undertaken by the Education Committee of the American Statistical Association a decade ago (Aliaga et al. 2007) but is currently under revision. Like the AMATYC, and the MAA's Curriculum Renewal Across the First Two Years (CRAFTY) subcommittee of CUPM and other CUPM documents, this work will present an accurate picture of the discipline and, at the same time, will show how it is not serving its students well in many institutions. Although the recommendations published in 2007 still stand, many departments of mathematics have not kept pace with the changes led by applied statistics over the past decade.

The set of recommendations for an introductory course discussed in Aliaga and colleagues (2007) were outlined by George Cobb (1992) and remain as important today as they were then, but the changes that they represent are now at an even greater distance from the situation discussed in the 2007 document:

- Emphasize statistical literacy and develop statistical thinking.
- Use real data.
- Stress conceptual understanding, rather than mere knowledge of procedures.
- Foster active learning in the classroom.
- Use technology for developing conceptual understanding and analyzing data.
- Use assessments to improve and evaluate student learning.

With these as a start, the *Guidelines for Assessment and Instruction in Statistics Education: College Report* (Aliaga et al. 2007) provides sample questions, activities, investigations, and assessments, which, along with the discussion of the previous recommendations, offer a framework for the introductory statistics course for general education.

From 1995 to the present, the work done at the two- and four-year college levels, as well as at the university level, supports and extends the work started by the MAA's

Committee on the Undergraduate Program in Mathematics in the mid-1960s and continuing through to the new undergraduate guidelines of 2015. Undergraduate mathematics departments have become aware that high-quality content alone is not sufficient to produce high-quality learning. A major portion of what students need to learn and be able to use resides in the interaction of that content with the ways in which the students have learned it. Departments from the community college level through graduate schools have realized that the goals must change to include high-quality curricula and programs grounded in learning and realistic, data-filled problems. Further, national scientific organizations, state boards of higher education, and professional groups are stepping forward to ensure that resources are available to provide both ongoing professional development for collegiate faculty and support for continued curricular development work (Snook 2004).

Chapter 3: The Implemented Curriculum

Historically, the implemented curriculum in school mathematics has been dictated by a mixture of individual state expectations for the topics to be taught within mathematics classes at various levels of education and, in some districts, associated student abilities and expected outcomes. Varying by state, these expectations have sometimes been established at the state level and, occasionally, at local school district levels. As a result, the quality of mathematics education received by a given U.S. student has historically been determined in large part by the quality of expectations set by state and local school authorities. The instructional materials selected by the state, the school district, or the individual schools have also set bounds on these expectations. Finally, the content and representations of it have been heavily influenced by the individual classroom teacher's mathematical knowledge and pedagogical knowledge. The mathematical standards process outlined in the previous two chapters of this report was developed precisely to address this confused mixture of expectations and delivery. Many in the broad mathematical sciences community believed that a standards-based guide to teachers' pre-service mathematics education, continued professional development in both mathematical content knowledge and related pedagogical methods, and an understanding of the intended curriculum would provide a more coherent delivery of that curriculum and an improved set of student outcomes.

The Impact of NCLB, ESSA, and CCSSM on the Implemented Curriculum

President George W. Bush's education plan, which his administration called No Child Left Behind (Bush 2001), outlined a program to establish state-level expectations and related state assessments in mathematics given annually to students in grades 3–8 and at one grade level in high school. By the time the No Child Left Behind Act of 2001 (NCLB 2002) was signed into law in early 2002, the foregoing mandates for states and their schools were the nation's law, as well as the basis for the development of an assessment-reporting structure requiring schools to meet annual student achievement progress goals known as Adequate Yearly Progress (AYP), beginning in the school year 2005–6 (NCLB 2002). The implementation of these global mandates took a variety of forms across the 50 states and the District of Columbia (Reys 2006; Smith 2010). One might say that variability was the constant in the implementation of state standards and their assessment by 2015.

The Every Child Succeeds Act (ESSA), the reauthorization of the NCLB act, was passed by Congress on December 9, 2015, and moved to President Obama for quick approval (Shober 2015). This act changes many of the regulations contained in NCLB, including the AYP requirements. At this writing, it is too early to tell what the Every Student Succeeds Act (ESSA) will mean for mathematics education. Removing the AYP regulation and permitting the states once again to set their own standards for proficient performance may, on the one hand, allow some relaxation of standards and a move of more states away from CCSSM. Or, on the other hand, states and schools may continue on the path set by CCSSM reforms.

In either case, the implemented mathematics curriculum in U.S. schools remains dictated to a large degree by the contents of the textbooks or other instructional materials, the assessments used locally and by state governments, the sequencing of the topics in those materials, and the overall alignment of the materials and the assessments with the standards in place. There appears to be greater consistency in what is expected, especially within textbook adoption states, as schools nationwide are beginning to work toward helping their students attain essentially the same set of learning outcomes. At present, textbooks based on the CCSSM curriculum are becoming available, but their influence is as yet unknown.

The National Assessment of Educational Progress does not routinely collect data on textbook usage, nor do other government sponsored educational surveys (Reys 2006). However, some data are available from publishers, assessment organizations, and isolated research studies. One independent research firm, Education Market Research, has conducted surveys for several years, and information from its most recent survey is used in the grade-level discussions that follow (Resnick and Sanislo 2015).

Anecdotal information suggests that three modes of operation compete for the role of major source for "instructional guidance" in the classroom. The first is the adoption of a single textbook for each grade level as the basal resource. This mode seems to be experiencing a downward trend, as the resources that school systems, schools, and teachers rely on are becoming more multifaceted. Although the majority of systems still appear to be using the single textbook approach to curricular resource materials, replacement options often include two textbooks for a grade or a textbook plus replacement chapters from another source for a couple of major curriculum topics. A second mode is school system–sponsored, teacher-written materials produced by the staff of a school system itself. These materials are used in different ways and sometimes are supplemented by digital computer-based practice software. A third source is full-blown digital curricula, presenting learning opportunities to a student at a computer terminal, implemented with or without class instruction or discussion. Publishers have also developed digital guided software that students can use along with their textbooks in a semi-independent fashion.

Solid data on the percentages represented by each of these three modes do not yet exist. Part of this variation in instructional material selection comes from the fact that CCSSM came on the scene as a proxy for a national curriculum very rapidly as compared with the longer, slower evolutionary path of the NCTM Standards movement. At this point, no track record exists for the use of specific materials based on CCSSM in the nation's classrooms. It is clear that digital instructional materials are playing an increasing role in most classrooms, ranging from supplementary practice materials to interactive discovery materials to complete interactive programs.

| Mathematics Materials in Elementary Schools (K–Grade 5) | An elementary school teacher who teaches reading, science, and social studies and is with the same students almost the entire day almost always also teaches these students mathematics. Teachers in K– grade 5 do not have time to create lessons for all these subjects, and as a result, they tend to rely heavily on the mathematics instructional materials purchased by their school district. Central to these is the role of the teacher's edition of the mathematics textbook. According to an industry survey, 78.8% of U.S. schools in 2014–15 reported using a basal mathematics series that they either follow very closely (41.7%) or from which they pick and choose (37.1%) as needed. The combined 78.8% using the textbook at least as a source is significantly lower than the combined percentage for the same two questions in prior years: 94% in 2001, 93.4% in 2005, and 88.1% in 2011 (Resnick and Sanislo 2015). These data clearly mark a pattern of decreased dependence on a single text series for guidance in what to do in mathematics class. |

In the past, most publishers marketed a coordinated elementary school curricular series to cover all grades, K–8. Beginning in the early 1980s, these series were split into two parts, a primary series for the K–5 or K–6 elementary school portion and a sometimes different series for the grades 6–8. More recently, publishers have segmented many of the K–5 portions of these programs into blocks covering K–grade 2 and grades 3–5 for marketing purposes, to parallel NCTM's *Principles and Standards* (2000).

Any discussion of the most commonly used textbooks must be written and read with great care because textbooks remain in use for several years after their initial purchases. Much of the everyday conversation about what is most popular focuses on what textbooks are currently being adopted. As a result, two distinct perceptions emerge regarding which textbooks U.S. students use. We will base our discussion of the status of textbooks on a national survey of textbooks in service in the school year 2014–15 (Resnick and Sanislo 2015). Data on textbook use indicated that mathematics teachers in K–grade 2 tend to stay with their textual materials for 3.6 years, on average. The three most frequently used series in K–grade 2 in 2014–15 were the following:

- *EnVision Math* from Scott-Foresman/Pearson (31.5%)
- *Everyday Mathematics* from Everyday Learning/McGraw-Hill (25.9%)
- *Investigations in Number, Data & Space* from Scott-Foresman/Pearson (13.0%)

Collectively, these three series accounted for slightly more than 57% of the series textbooks in use in K–grade 2 in 2014–15.

The data on textbooks in grades 3–5 suggest that teachers at this level tend to be using adopted materials for a longer period of time than teachers in almost all the other curriculum domains. Teachers of mathematics in grades 3–5 have been using their present textbook for an average of 4.5 years. The following were the three most frequently used series in grades 3–5 in 2014–15:

- *EnVision Math* from Scott-Foresman/Pearson (23.6%)
- *Everyday Mathematics* from Everyday Learning/McGraw-Hill (12.7%)
- *Math Expressions* from Houghton-Mifflin/Harcourt (9.1%)

The three series accounted for slightly more than 45% of the textbooks in use at these grade levels (Resnick and Sanislo 2015). Note that reports of the textbooks being used most frequently do not provide an account of the entrance of newer materials based on CCSSM, since the textbooks named only account for essentially 45% of the materials used at these grade levels.

Mathematics Materials in Middle or Junior High Schools (Grades 6–8)

The patterns of use are much more difficult to summarize in an analysis of the data for grades 6–8 because many districts use a mixture of a basal series that covers the entire span of content topics for some of the grades but select other textbooks covering algebraic content at either an introductory level or the level that would typically be found in the first years of high school (grades 9–10). In addition, approximately 5% of the grade-level enrollment is taking a course in geometry. Mathematics teachers at these grade levels are teaching out of a textbook that they have used, on average, for about 3.6 years. Again, this data point gives some distributional information but neglects to note it as a mark of centrality in a distribution from cases of "just-starting-to-use" to cases in which the book has been used for an eternity.

With respect to the textual materials currently in use in 2014–15, data were reported on the basal 6–8 textbooks, pre-algebra materials, and algebra 1/geometry textbooks. For the basal series, the three leading series were the following:

- *Holt McDougal Mathematics* from Holt McDougal/HMH (10.9%)
- *Connected Math* from Prentice Hall/Pearson (9.8%)
- *Math Connects* from Glencoe/McGraw-Hill (7.6%)

These three series were found to account for 28.3% of textbooks used for these grade levels. Other textbooks mentioned were single texts, not series. These included pre-algebra textbooks from Glencoe/McGraw-Hill and Prentice Hall/Pearson. These textbooks accounted for another 20.1% of textbooks used in schools at grades 6–8. Two textbooks were tied in being mentioned for the algebra 1 courses in grades 6–8. These textbooks were *Algebra I* from McDougal Littell/HMH (4.3%) and *Algebra I* from Prentice Hall/Pearson (4.3%).

Mathematics Materials in Senior High Schools (Grades 9–12)

High school mathematics programs, like programs at the other levels, seem to be on a seven- to eight-year rotation program for consideration of textual material replacement. The average adoption lengths at grade 9–12 are essentially normally distributed around a mean of 3.8 years. At the high school level, the mainstream core curriculum currently found in U.S. secondary school classrooms is built around a sequence of three full-year courses—algebra 1–geometry–algebra 2, or algebra 1–algebra 2–geometry—beginning in eighth, ninth, or tenth grade, followed by a fourth year of precalculus, usually giving strong attention to functions and trigonometry. Since the mid-1950s, an increasing percentage of students have completed a year of calculus at the high school level. This latter course, especially when it is an Advanced Placement Calculus course, usually covers the content ordinarily found in the first semester of university-level calculus. In about 20% of these cases, this course covers the equivalent of the first full year of university-level calculus. In most school districts where students participate in AP Calculus courses, algebra 1 is taught in the eighth grade.

More than 90% of the secondary schools in the country follow the more traditional course-based curriculum, rather than an integrated curriculum, for the majority of their students. Many of them present it at a slower pace for lower-performing students, subdividing some of the courses over two years or altering the number of topics covered in a course. Now many school districts are presenting the courses for lower-performing students as double-period classes, giving these students for whom time on task was the issue a chance to keep pace with the curriculum across their time in high school. Since 1990, several integrated secondary school mathematics curricula have been developed. These programs blend the content of algebra, geometry, functions, and data analysis in a highly connected and integrated fashion, usually with an emphasis on modeling and applications (Hirsch 2007). Data on the percentage of students studying the integrated curricula are hard to obtain and verify, but estimates commonly place it between 4% and 5% of secondary students. Only one integrated series was used in at least 6% of U.S. schools in 2014–15. This was the *College Preparatory Mathematics Program* from CMP Educational Program.

When the focus was on single textbooks used in more than 6% of schools, four texts emerged from the data, two at the algebra 1 level and one at the algebra 2 level, and one geometry text. These four textbooks and the percentage of schools that they accounted for were three algebra and one geometry book, tied in two groups of two: *Algebra I* from McDougal Littell/HMH (Larson) and *Algebra I* from Prentice Hall/Pearson, both with 9.7% of the schools. The final two individual textbooks were *Geometry* from Prentice Hall/Pearson in 6.5% of the schools and *Algebra II* from Prentice Hall/Pearson in 6.5% of the schools. The number of separate courses and differing course names makes any finer analyses of the reach of different textbooks difficult (Resnick and Sanislo 2015). None of the high school series developed with initial support from the NSF has the breadth of use that is enjoyed by the series' elementary and middle school counterparts. Together with the NCTM Standards documents, these NSF project-related series have influenced mainstream texts, motivating publishers to include more applications and more work with technology. At the same time, pressure from colleges has influenced the publisher of these texts to maintain, if not increase, both conceptual and procedural skill work with algebra and functions. At the middle school level, the Pearson text series *Connected Mathematics* is a curriculum development project that had its start in one of the NSF-funded middle-school projects developed to build curricula based on the NCTM Standards. Other series have garnered shares of the market, but none of the high school or other middle-school NSF projects has had the influence or wide pattern of adoptions that the *Connected Mathematics* series has had.

The Use of Digital Media in the K–12 Classroom

Educational Market Research carried out surveys of the adoption of digital media in the nation's schools from 2003 to 2014 as part of its annual collection of supplementary products to printed textbooks. Whereas in the past most of the companies surveyed in the annual surveys were publishers or manufacturers of instructional materials, a sharp change in media used for delivering supplementary materials occurred in the K–12 market from 2009 to 2011. For this period, examination of the relative responses for print delivery usage and online/digital delivery shows the former trending upward from 61.4% usage in 2009 to 71.3% usage in 2011, and modulating back down to 65.2% in 2014. Overall, the best that one can say is that the use of print media to deliver supplementary instruction has held steady.

A comparison of the percentages of sales of educational products reported in four categories for the years 2013 and 2014 showed an increase from 34.6% to 37.2%, respectively, for digital media. Print instructional materials decreased from 46.4% to 43.8% over the same period. In the other two categories of products reported, hardware/furniture/equipment remained stable at 15.2%, as did "other" at 3.8%. In this part of the survey, data related to calculators appeared under "equipment." The data point reported was a 12.6% decrease in year-to-year sales attributable to calculators from 2012 to 2013. This is a single data point, but it may indicate a change in priorities from calculators to other forms of digital media in mathematics classrooms (Resnick and Sanislo 2015).

Although these data reflect overall disciplinary sales, the responses of mathematics administrators, secondary mathematics teachers, and elementary mathematics teachers added meaning to similar data. In 2011, the most frequent response regarding the percentage of time spent on the use of digital materials in mathematics was less than 5% of class time. By 2013, the most frequent response was 11% to 25% of the time, with no one selecting less time than this category of response. When asked for the names of websites that they frequently used, teachers by grade-level intervals gave the following responses (limited to 10% or more):

- K–grade 2: IXL.com (22.8%)s, CoolMath.com (10.5%), MathPlayground.com (10.5%)
- Grades 3–5: CoolMath.com (25.5%) and IXL.com (23.5%)
- Grades 6–8: KhanAcademy.com (24.8%), publisher material sites (19.5%), IXL.com (18.6%), and StudyIsland.com (12.4%)
- Grades 9–12: KahnAcademy.com (50.8)%, publisher sites (18.5%), and PurpleMath.com (10.8%)

The number one factor that teachers responded to in 2013 as a criterion that they used in selecting digital media was the alignment of the materials with CCSSM. The next two most important factors were emphasis on real-life applications and problems (60.6%) and usefulness in integrating basic skills and problem solving (42.4%) (Resnick and Sanislo 2015).

Coursework and Pedagogical Patterns in K–Grade 12 Mathematics

During the past twenty-five years, high school graduation requirements and college admission requirements have increased, along with the percentage of four-year colleges and universities now requiring two years of algebra and a year of geometry for admission. Also, as a result of the impact of technology on everyday lives and worries about U.S. students' lackluster mathematical performances in international studies, the public appears to have become more aware of the role that mathematics can play in the future lives and careers of secondary school students. Furthermore, as the data in tables 2 and 3 indicate, high school mathematics enrollment data, when combined with course-taking data from the middle grades, show that more students are taking courses in algebra before high school. This change has in turn contributed to a steady and significant increase over time in the percentage of students topping out at a higher level in the mathematics curriculum than had been reached by students in previous years.

The data in table 2 reflect the shifts in the transition of students from the regular eighth-grade curriculum to a pre-algebra class and then to an algebra 1 class over time. The bulk of the data for "Other course" at present is enrollment in the equivalent of a high school geometry course. So when we see the decline in enrollment in high school courses at the level of algebra 1 or below, we are seeing an increase in the percentage of secondary school students enrolling in at least geometry or algebra 2 as their most advanced mathematics course taken in high school, having taken the lower-level courses at the middle school/junior-high school level.

The data in table 2 are from the NAEP Long-Term Trend Studies, a series of survey assessments using basically the same instruments and questionnaires over time. As a result, the data on course taking in U.S. schools are some of the most reliable, with the questions about courses and the framework of courses having remained very constant over time. Unfortunately, one problem with the data for the secondary-level courses is that they merge the percentage of students taking precalculus with that of students who might be indicating calculus as their highest course taken.

Table 3 contains data from the main NAEP mathematics assessment, an assessment that is given on a different schedule and for which the survey questions, as well as the test items, are revised more frequently. There are differences in the total numbers ascribed to the percentages of students taking precalculus or calculus in these two different NAEP assessment programs. However, the ratio between precalculus enrollments

Table 2

Percentage of students by most advanced mathematics course taken in middle school and taken in high school

Middle school course / year	1986	1990	1994	1999	2004	2008	2012
Reg. grade 8 math	61	57	43	37	31	31	28
Pre-algebra	19	23	32	34	32	32	29
Algebra 1	16	15	20	22	29	30	34
Other course	5	5	4	6	6	7	9
High school course / year	1986	1990	1994	1999	2004	2008	2012
Pre-algebra or general math	18	15	9	7	4	3	2
Algebra 1	18	15	15	11	9	7	5
Geometry	16	15	15	16	16	17	15
Algebra 2/Trigonometry	40	44	47	51	53	52	54
Precalculus or Calculus	7	8	13	13	17	19	23

Percentage of students by the most advanced course they were enrolled in during their final year of middle school and secondary school, NAEP Long-Term Trend data from the studies in the years shown (Mullis et al. 1991a; Campbell, Voelkl, and Donahue 1997; Campbell, Hombo, and Mazzeo 2000; Perie, Moran, and Lutkus 2005; Shettle et al. 2007; Nord et al. 2011; NCES 2013a).

and calculus enrollments in Main NAEP decreased from 2.25:1 to 1.06:1 and increased back to 1.44:1 over the period of time from 1990 to 2013 ("Main NAEP" is discussed at the beginning of Chapter 4). This pattern follows the data pattern seen in table 2, with a gradual increase in the selection of more advanced mathematics courses as part of a student's high school courses. It may be that all who aspired to higher courses were not as capable as those who originally chose that path in earlier years.

Table 3

Percentage of students taking precalculus or calculus as their most advanced class

Course	1990	1996	2000	2005	2009	2013
Precalculus	9	14	17	21	24	26
Calculus	4	7	16	18	18	18

Breakdown of percent data for students enrolled in precalculus or above in data from Main NAEP in the years shown (Mullis et al. 1991b; Reese et al. 1997; Braswell et al. 2001; Lee, Grigg, and Dion 2007; NCES 2009b, 2013d).

Confirming these data, Blank, Langesen, and Peterman (2007) report that the percentage of high school students completing algebra 2 increased 13% from 2000 to 2009, and the percentage completing precalculus increased 24% over the same period. Concurrent with these changes has been an increase in the numbers of students taking Advanced Placement courses in mathematics (Shettle et al. 2007; College Board 2011; Nord et al. 2011; Blair, Kirkman, and Maxwell 2013).

Prior to the 1990 shift of the use of the National Assessment of Educational Progress (NAEP) to collect information for state comparisons and for demographics-based monitoring of students' achievement in the nation's schools, greater attention was given to the collection of information on curriculum and instructional variables in the nation's class-rooms. Starting with the 2005 assessment, these data were severely reduced, and many of the long-term lines of data were truncated.

NAEP has tracked the use of calculators at the classroom level since the 1980s. Initial data showed that schools owned sets of calculators for instructional purposes. But by 2005, the data indicated that 76% of the nation's grade 4 students reported owning a regular calculator, and 6% reported owning a graphing calculator (NCES 2011). More recent NAEP assessments have stopped asking this direct question, shifting more attention and questions to the use of computers.

In 2005 through 2015, fourth-grade teachers were asked about the levels of calculator use that they allowed their students in taking a mathematics test or quiz. Table 4 contains the responses in terms of the percentage of grade 4 students falling into each usage class and the mean score scale for students in that usage class. The results show a slight, but statistically significant, decline initially in the percentage of students reported as never using a calculator and then a statistically significant increase in the number said never to be allowed the use of the calculator. The data for sometimes allowing the use of calculators on assessments has also declined. The data suggest that students using a calculator score marginally higher than those never using one.

Table 4

Percentages and related scale scores of fourth-grade students by their teachers' answers to the question, "When you give students a test or quiz, how often do they use a calculator?" Data from Main NAEP Assessments of the years shown.

Usage	2005		2007		2009		2011		2013		2015	
	Percent	Scale score	Percent	Scale score	Percent	Scale score	Percent	Scale score	Percent	Scale score	Percent	Scale score
Never	75	238	67	240	71	240	72	240	77	241	82	241
Sometimes	25	240	32	241	28	243	27	243	22	243	18	242
Always	0	dna	1	231	1	231	2	232	1	232	#	226

– Rounds to zero

Breakdown of percent data for students enrolled in fourth grade, using data from Main NAEP in the years shown (Perie, Grigg, and Dion 2005; Grigg, Donahue, and Dion 2007; NCES 2009a, 2011, 2013b, 2015c).

At the grade 8 level, teachers were asked about the type of calculators, if any, that their students used during mathematics lessons. The results from the NAEP assessments of 2009 to 2015 showed the data reported in table 5. Here the data are very stable, with teachers reporting that 72% to 73% of the students were using either a scientific calculator without graphing capabilities or a scientific calculator with graphing capabilities. An analysis of these data for both years also shows significant differences in students' NAEP achievement scores as the technology changed from the basic four-function to a scientific (not graphing) calculator and then again to a graphing calculator. Also interesting is the fact that the use of a basic four-function calculator at the grade 8 level did not give a student an advantage over a student without the use of a calculator with respect to their scale scores. It may be that at this juncture knowing one's facts makes up for not knowing how to use, or having access to, the basic calculator.

Table 5

Percentages and related scale scores of eighth-grade students by their teachers' answers to the question, "What kind of calculators do your students use in mathematics lessons?" Data from Main NAEP Assessments of the years shown.

Type of calculator	2009		2011		2013		2015	
	Percent	Scale score	Percent	Scale score	Percent	Scale score	Percent	Scale score
None	10	273	11	275	11	277	5	279
Basic four function (+, −, ×, ÷)	17	273	17	274	15	274	14	271
Scientific (not graphing)	48	286	46	286	47	287	53	284
Graphing	25	290	26	291	26	292	28	288

Breakdown of percent data for students enrolled in fourth grade, using data from Main NAEP in the years shown (NCES 2009a, 2011, 2013b, 2015c).

Mathematics Study at the Postsecondary Level

At the postsecondary level, students have a wide variety of options for studying mathematics. Coursework is available through community colleges, universities, and a variety of vocational schools, work-based educational programs, and commercial outlets. The data collected every five years by the Conference Board of the Mathematical Sciences (CBMS) provide the best trend data for curricular programs and enrollments in two- and four-year colleges. Unfortunately, the 2015 report was still in the process of data collection and analysis at the time of the writing of this report. Hence, the data used in the following is an amalgam of data from the CBMS study (Blair, Kirkman, and Maxwell 2013) and the *Digest of Education Statistics 2013* (Snyder and Dillow 2015).

Mathematics courses at the types of postsecondary institutions mentioned above range from arithmetic and pre-algebra to linear algebra and differential equations at vocational and two-year colleges, and from intermediate algebra and precalculus through advanced graduate courses at four-year institutions and universities. Tables 6 and 7 demonstrate this wide range and the change in mathematics enrollments from 1980 to 2010 at two- and four-year colleges, respectively. In these tables, remedial courses include arithmetic, pre-algebra, and elementary and intermediate algebra. Note that remedial course enrollment is correlated with a lower probability of graduation.

Precalculus courses include college algebra and trigonometry as well as finite mathematics, non-calculus-based business mathematics, mathematics for prospective elementary school teachers, and other courses for non-science majors. Calculus includes both mainstream and non-mainstream courses (e.g., calculus courses tailored to students in other majors, such as life sciences or business). These tables do not include mathematics courses taught outside mathematics and statistics departments. Enrollments are for the fall quarter or semester of the 2010–11 academic year (Blair, Kirkman, and Maxwell 2013).

Two-year college enrollments increased over this same period and were rising from the nearly 6.18 million enrolled in 2005 to slightly more 7.22 million in the fall of 2010, an increase of about 16.7% (Snyder and Dillow 2015; Hussar and Bailey 2013). An examination of the data in table 6 shows that this same period saw an increase of about 18.9% in the number of students enrolled in mathematics. The National Center of Educational Statistics projection for two-year college enrollment in 2015 and 2020 are 7.63 million and 8.21 million, respectively (Hussar and Bailey 2013).

Not only did enrollments increase overall, but the increase also occurred across the full range of the two-year college offerings. Remedial enrollments were up by 19.3%; precalculus enrollments, by 13.7%; calculus enrollments, by 29.0%; statistics enrollments, by 16.1%; and enrollments in other courses (liberal arts, math for elementary teachers, and so on), by 21.5%. This pattern contrasts with four-year college data over the same period, as the data in table 6 show.

Table 6

Estimated enrollment (in thousands) in mathematics courses in two-year colleges

Course	Year						
	1980	**1985**	**1990**	**1995**	**2000**	**2005***	**2010***
Remedial	441	482	724	800	763	964	1150
Precalculus	180	188	245	295	274	321	365
Calculus	86	97	128	129	106	107	138
Statistics	28	36	54	72	74	118	137
Other	218	133	144	160	130	186	226
Total	953	936	1295	1456	1347	1696	2016

*Data in 2005 and forward reported by sections by average size rather than by percentage of total students calculations. From Blair, Kirkman, and Maxwell (2013).

Table 6 shows that from 1985 to 2010, more than half of the mathematics enrollments in two-year colleges have been at the remedial level. The overall increase in the number of mathematics courses in two-year colleges is partially a function of the overall increase in enrollments at these institutions. The increase is also, however, partially a function of the increased realization that mathematics enables knowledge and opportunity (Lutzer et al. 2007; Blair, Kirkman and Maxwell 2013).

Four-year college enrollments in mathematics increased over this same period of time from the nearly 11.0 million enrolled in the fall of 2005 to slightly more than 13.3 million enrolled in the fall of 2010, an increase of about 21.2% (Snyder and Dillow 2015; Hussar and Bailey 2013). An examination of the data in table 7 shows that this same period saw an increase of about 22.5% in the number of four-year college students enrolled in mathematics. The National Center of Educational Statistics projection for four-year college enrollments in 2015 and 2020 are about 13.6 million and 14.8 million, respectively (Hussar and Bailey 2013).

Not only did enrollments increase overall between 2005 and 2010 but the increase also occurred across the full range of the four-year college offerings. Remedial enrollments were up by 4%, showing a reverse in percentage of change over the last two survey periods, precalculus enrollments were up by 22.2%, calculus enrollments were up by 30.3%, and statistics enrollments were up by 44.5%. Enrollments in advanced mathematics courses were up by 33.9%. This pattern continues the increasing demand for these courses seen in the last three surveys and marks a decided increase over the last survey at the same time.

Table 7

Estimated enrollment (in thousands) in undergraduate mathematics and statistics courses in four-year colleges

Course	Year						
	1980	1985	1990	1995	2000	2005*	2010*
Remedial	242	251	261	222	219	201	209
Precalculus	602	593	592	613	723	706	863
Calculus	590	637	647	538	570	587	765
Statistics	dna	dna	125	143	171	182	263
Other	91	138	119	96	102	112	150
Total	1525	1619	1744	1612	1785	1788	2250

*Data in 2005 and forward reported by sections by average size rather than by percentage of total students calculations. From Blair, Kirkman, and Maxwell (2013).

The graph in figure 2 shows the consistent growth in enrollment for both two-year and four-year colleges and their contributions to the total number of undergraduate students enrolled in mathematics at U.S. two-year colleges and four-year colleges and universities. Although differences occur in the rates of growth of individual subareas within each subdivision of postsecondary education, one can see the increasing percentage of the total contributed by the two-year college enrollments in mathematics over time.

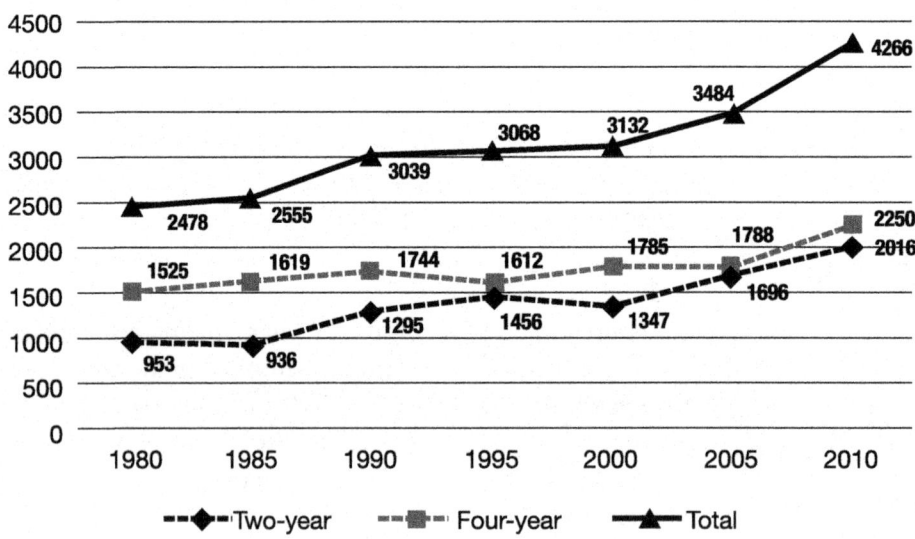

Fig. 2. Undergraduate enrollments (in thousands) in mathematics 1980–2010

Changes in the Number of Baccalaureate Degrees in Mathematics

Data from the American Freshman study indicate the changes by year in the percentage of freshmen entering baccalaureate granting institutions with an intention to major in mathematics or statistics: 0.9% (2010), 0.9% (2011), 0.9% (2012), 1.0% (2013), and 1.1% (2014) (Higher Education Research Institute 2010, 2011, 2012, 2013, and 2014; National Science Board 2014). These percentages are significantly lower than those that were observed in the 1960s. Although the number of bachelor's degrees in mathematics is lowest among the areas listed in the sciences by the National Center for Science and Engineering Statistics, the mathematics requirements of majors outside the physical sciences, however, have increased significantly in the same period. Although some of the mathematics needed to fulfill these requirements is taught outside departments of mathematics and statistics, the increases in these requirements are a major factor in the overall increase in the number of courses taken in departments of mathematics and statistics (National Science Board 2014). At the same time, the piecemeal taking of courses to fulfill such demands does not imply a ready, long-term supply of mathematically trained individuals to meet the nation's needs.

Chapter 4: The Attained Curriculum

Central to the measure of the success of a curriculum is the academic attainment of the students who have participated in the instructional experiences associated with it. Unfortunately, outside of test-item data from the mathematics tests administered through the National Assessment of Educational Progress, little information is available at a national level on student progress in the diverse curricula offered throughout the United States. As a result, we examine the attained curriculum through data from NAEP. At the time of this writing, initial data from the first administrations of assessments developed by the two consortia to provide information on student achievement to the states adopting the Common Core State Standards for Mathematics have not provided a detailed set of data for comparisons. Outside of NAEP, the national and state-level data from the two major college entrance examinations—the ACT and the SAT programs—provide other stable assessments of student achievement outcomes. These are not limited to the flagship examinations but extend to the various subject-matter examinations provided by the two programs, as well as their programs marking progress toward college and the workplace.

The National Assessment of Educational Progress (NAEP)	The U.S. government, through the Department of Education's National Center for Education Statistics (NCES) and guidance from the National Assessment Governing Board (NAGB), administers a large-scale assessment program under the title of the National Assessment of Educational Progress (NAEP). This program's mathematics assessments began in 1973 and continue to be administered periodically to assess student knowledge of, and opportunity to learn, mathematics by surveying random samples of American youth. This survey of the mathematical abilities of U.S. students has taken on even more importance with the use of NAEP as a barometer for measuring states' performances relative to the strictures of the NCLB legislation.

When people in the United States mention NAEP, they are usually referring to what is known as Main NAEP. This is the program that focuses its assessment on a random sample of students from grades 4, 8, and 12 according to a schedule of recurring assessments beginning in 1973. The current program for Main NAEP assessments is every two years on the odd years for grades 4 and 8 and every four years for grade 12 on the odd years beginning in 2017 (NAGB 2015).

In addition to each Main NAEP assessment since 1990, a random sample of students from each state has also been assessed to provide each state with the same data as Main NAEP provides to the nation. This program is called State NAEP, and it has been in place since 1990. A third NAEP program, known as TUDA NAEP, is the Trial Urban District Assessment, which began in 2002. This program provides a NAEP assessment profile to the nation's largest urban districts at grades 4 and 8 through expanded samples of students drawn in conjunction with the State NAEP assessments. This program is limited to large urban school districts with a minimum of roughly 20,000 students in grades K–12 and is given in the same years as Main NAEP.

The final NAEP program is the NAEP Long-Term Trend Assessment (LTTA). This assessment differs from the other NAEP assessments in that it selects random samples of 9-, 13-, and 17-year-olds, thus maintaining the practice followed over time in this trend analysis. The last LTT assessment was conducted in 2012, and the next one will be in 2024. The test for the Long-Term Trend Assessment in mathematics has remained relatively unchanged over time since its first administration in 1973. Consequently, the LTTA results provide a barometer for measuring today's students' achievement against the expected performance of previous generations—that is, these data help to answer the question, Can students today still do what their parents were taught and expected to do? Minor changes have been made to the LTTA over its lifespan, but in each case statistical and validity studies have been carried out to ensure that the goals of the assessment and the reporting levels do not require a break in the trend reporting (NAGB 2015).

We look at the most recent results of each of the assessments in the following sections, as well as at the trends over time for each group of students.

Main and State NAEP	The formats and expectations of Main and State NAEP Assessments are released a few years ahead of a given assessment in a NAEP Mathematics Framework. Frameworks for recent and present programs can be obtained from the NAGB website (NAGB 2015). The assessments are developed by a committee consisting of teachers from the grade levels assessed, collegiate mathematics and mathematics education professionals, and test and measurement experts from NAGB and the firms contracted to carry out the assessments in the field. The assessment itself is presented in a balanced incomplete block design, starting with blocks of items assembled into test booklets, each consisting of a fixed number of blocks of items, as well as a set of student questions concerning academic experiences and demographics. Over time, individual blocks of items are released, and new blocks of items are inserted to keep the test focused on the content targets enunciated in the current NAEP framework. The NAEP assessment in mathematics is focused on measuring the implemented curriculum, not on research into what curricular experts are thinking might be appropriate for a given grade level. It is a test of what is currently being taught in U.S. classrooms, not what might be taught.

Changes in NAEP Frameworks for Grades 4 and 8 Mathematics for 2010–15	The content frameworks for the NAEP mathematics assessments have changed in recent years to give more emphasis to algebra and function concepts and their applications. They also focus more attention on students' proficiencies in context-based problem solving; in constructing and communicating their own responses; and in knowing when and how to apply technology, where it is allowed, to solve problems on NAEP assessments. The 2009 and 2011 Main NAEP mathematics frameworks reflected changes taking place in expected outcomes for students in grade 12. Most of these were related to additional questions regarding topics in high school geometry and second-year algebra courses (NAGB 2010).

A current concern across all grade levels for Main NAEP is the rapid movement to the Common Core State Standards, a revolutionary change to common standards across the 43 adopting states and the District of Columbia. These standards differ significantly (Daro, Hughes, and Stancavage 2015; Hughes et al. 2013; Kane 2015) from the curriculum that has been in place for essentially the previous 20 years—since the time when state standards changed to become some version of the NCTM Standards as delineated in *Curriculum and Evaluation Standards* (1989). The difference is not so great in the content itself, although there are content differences. The biggest issue is the specific grade placement of particular content. For example, CCSSM expects students to demonstrate procedural knowledge in number and operations later, at a higher grade level than recommended in the NCTM Standards. And CCSSM places less emphasis than the NCTM Standards on geometry and data analysis, probability, and statistics from the early grades through grade 8. The delaying of topics and the lessening of emphasis on them in the school classroom can, over a short period of time, lead to a discontinuity in the assessment results. But NAEP is supposed to be a test of what students have learned from the curriculum that they encounter in their classrooms. Each of the changes in the assessments makes evaluating the changes over years in student performance more difficult from a policy point of view.

Executive Summary for NAEP 2015 Grade 4 and Grade 8 Results

The two sections that follow provide an overview of the NAEP 2015 results in perspective for both grade 4 and for grade 8. The overviews are somewhat repetitive as they cover the assessments at the two grades, each in some detail. However, to meet the needs of individual readers, both grade levels are covered in the same degree of detail.

In 2015, about 7,810 schools (7,230 public and 380 private) and 139,900 students (134,700 public and 2,400 private) were involved in the grade 4 mathematics assessment, and 6,150 schools (5,670 public and 340 private) and 136,900 students (132,500 public and 2,300 private) participated in the grade 8 mathematics portion of the Main NAEP assessment. The fact that the public and private school numbers do not always sum to the totals within a category is due to rounding that occurs in processing and in the weightings assigned to the scores to meet the sampling design specifications. The students were randomly selected according to a complex sampling design to form the basis from which results at both national and state levels could be developed and compared statistically. In addition to students from public and private schools in the 50 states, students were also selected for the collecting of data to develop NAEP reports on student performances for the District of Columbia and the Department of Defense Education Activity (DoDEA) schools (NCES 2016).

NAEP 2015 Grade 4 Results

The national fourth-grade mean scale score in mathematics performance in 2015 was 240. This score was lower than that observed in the 2013 NAEP fourth-grade mathematics assessment (242) by 1 point when the data were not rounded to the nearest integer value but was significantly different statistically ($p < 0.05$) from the 2013 value. The educational significance of that difference is still under study, but the information in figure 3 illustrates that the grade 4 national average score has increased by 16 points on the NAEP scale since the inception of the current model for NAEP trend assessments in 1996. This model allows for accommodations for students qualifying for the trend assessment. The scores for 1990 and 1992 reflect a time when accommodations were not permitted for qualifying students. Reporting in NAEP is limited to reporting at the nearest tenth of a score point. However, statistical tests for significant differences are calculated by using unrounded average scores.

Because of the large sample sizes involved in the NAEP assessment for a given grade level, a small change in the mean performance may be judged as statistically significant, even when the actual difference in performance is less than getting one more item correct on a NAEP assessment. However, what one sees when looking at the mean performances for fourth graders over the set of assessments since 1996 is impressive steady

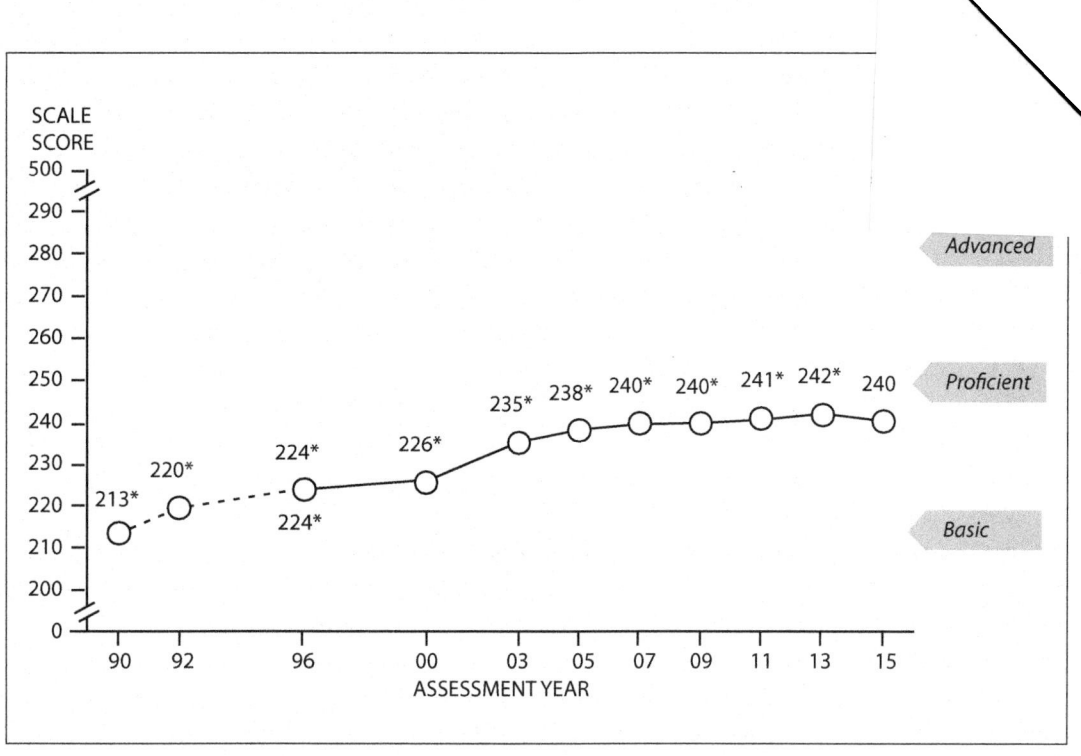

Fig. 3. Trend in fourth-grade NAEP mathematics mean scale scores. Graphic from the National Center for Education Statistics, *The Nation's Report Card: A First Look— Mathematics and Reading 2015*. Washington, D.C.: U.S. Department of Education, National Center for Education Statistics, 2015d. Used with permission.

growth in student performance until 2015. The values marked with an asterisk in figure 3 indicate years in which the national performance was statistically different from what was observed in 2015.

Achievement levels

A perusal of other national results observed for fourth graders in the Main NAEP 2015 mathematics assessment reveals an analysis of the percentage of 2015 students being classified in each of the NAEP mathematics achievement levels established in conjunction with the No Child Left Behind legislation in 2002. The data show that in 2015, the national sample of U.S. students could be described as consisting of 82% of students performing at or above the Basic level cut point. By complement, 18% of the nation's fourth graders were performing at a level below the Basic designation. Forty percent of the nation's fourth graders were performing at or above Proficient, and 7% of the nation's fourth-graders performed at the Advanced level.

The comparison of these values across the years in the decade from 2005 to 2015 is shown in figure 4. The figure gives the relative percentage of the nation's fourth graders within each of the achievement levels. The column of values at the right reflects the percentage of the fourth graders reaching or exceeding the Proficient level. This achievement comparison showed a significant decrease in the percentage of fourth graders reaching the targeted Proficient level in 2015 as compared with the 2013 percentage. However, there is no significant difference in the 2015 performance and that observed in 2011, 2009, and 2007. A significantly greater percentage of 2015 fourth-graders achieved the Proficient target level than students from the 1990 through 2005 cohorts.

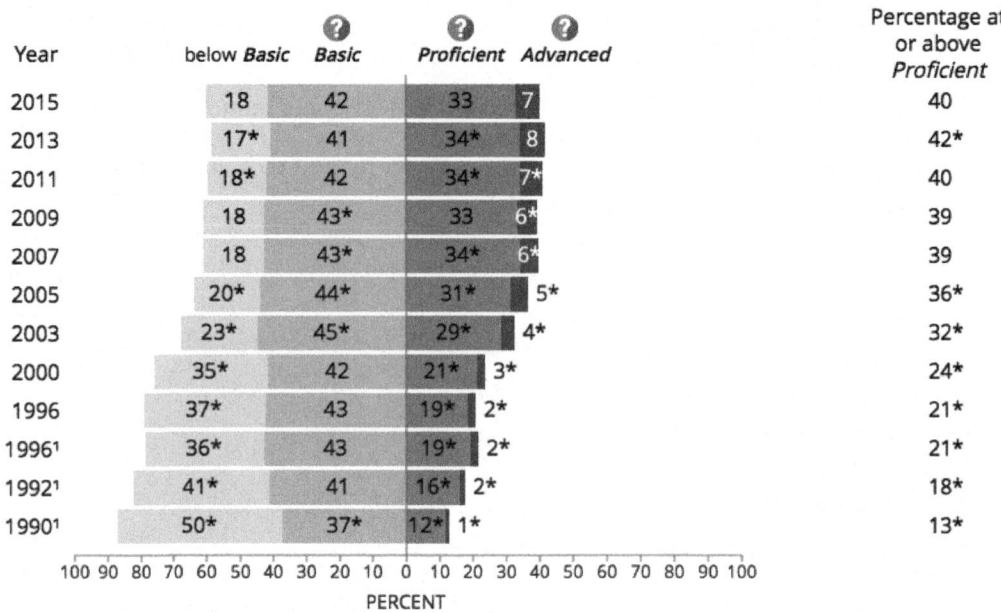

Year	below *Basic*	*Basic*	*Proficient*	*Advanced*	Percentage at or above *Proficient*
2015	18	42	33	7	40
2013	17*	41	34*	8	42*
2011	18*	42	34*	7*	40
2009	18	43*	33	6*	39
2007	18	43*	34*	6*	39
2005	20*	44*	31*	5*	36*
2003	23*	45*	29*	4*	32*
2000	35*	42	21*	3*	24*
1996	37*	43	19*	2*	21*
1996[1]	36*	43	19*	2*	21*
1992[1]	41*	41	16*	2*	18*
1990[1]	50*	37*	12*	1*	13*

100 90 80 70 60 50 40 30 20 10 0 10 20 30 40 50 60 70 80 90 100

PERCENT

Fig. 4. Percentage of grade 4 students in each NAEP achievement level by year. Graphic from the National Center for Education Statistics. *The Nation's Report Card: A First Look— Mathematics and Reading 2015*. Washington, D.C.: U.S. Department of Education, National Center for Education Statistics, 2015d. Used with permission.

The NCLB legislation's performance goals called for all students in the nation to be able to perform at the Proficient level by the spring of 2015. The data in figure 4 provide evidence for educators' comments in the early 2000s to the effect that the targeted NCLB goal for all students to reach the Proficient level by 2015 was nothing more than a pipe dream, one that neglected children's development and a myriad of other factors affecting school achievement in the United States. This and the AYP requirement based on the levels were both removed with the passage of the Every Student Succeeds Act in December 2015.

Gender comparisons

Comparisons for male and female fourth-grade mathematics performances in 2015 showed male fourth graders performing at a mean performance level of 241, whereas females were performing at 239. A statistical test showed this difference to be significant at the $p < 0.00001$ level. The results for the years from 2005 to 2015 show the differences between the two genders growing smaller but with the direction continuing to favor males across the decade. State-by-state results for grade 4 are discussed later in this chapter.

Race and ethnicity differences

Comparisons of racial and ethnic differences are loaded with a number of correlated factors. Given the inability to cover them in a completely scholarly way, we present the nature of the trends in each group over time and leave the discussion of the gaps between different groups to later publications. Data exist to provide a basis to talk about trends across the past decade, 2005–2015, for students of White, Black, Hispanic, and Asian/Pacific Islander descent. The groups' percentages of the 2015 student fourth-grade population in 2015 were 51%, 15%, 25%, and 6%, respectively. The remaining groups, for which full information was not available, were Asians (5%), American Indians/Alaska Natives (1%), and students classified as being of two or more races (3%).

White student performance began the decade with a mathematics assessment mean scale score of 246 for 2005, a score that represented a significant increase over the prior assessment in 2003. In 2007, the grade 4 performance of White students increased significantly, to 248. In 2009, the fourth-grade level remained the same for White students, at 248. In 2011, White fourth graders again improved significantly, with a score of 249. In 2013, White student performance reached 250, a level that was significantly greater ($p < 0.05$) than this group's performance in 2011. However, in 2015, White students' performance dropped to 248, a statistically significant decrease from their 2013 level. This drop contributed heavily to the national decline, owing to the percentage of the population that was White.

Black fourth graders started the decade in 2005 with a score of 220, a significant increase over their previous 2000 score. Black students' performance increased significantly, to 222 in 2007, and held at this 222 level, with no change, in 2009. In 2011, Black students' mean performance increased significantly to 224, a level that they held with no change through 2013 and 2015. As a result, the performance of Black students was not part of the decrease in overall fourth-grade performance in 2015.

Hispanic fourth graders began the decade at a mean of 226, a level that marked a statistically significant growth over their 2003 performance. Hispanic fourth graders' mean mathematics performance increased significantly, to 227 in 2007 and held steady at this level, with no change, in 2009. In 2011, Hispanic fourth graders performance increased significantly, to 229, followed by a non-significant change to 231 in 2013. In 2015, Hispanic fourth graders' performance fell back to 230. None of these latter changes were significantly different from the preceding level. Because of the large percentage of the fourth grade population that is Hispanic, this difference may account for some of the decrease in overall U.S. fourth-grade performance in 2015.

Asian/Pacific Islander students started the decade at 251, a level that was a statistically significant increase over their 2003 level. They then increased significantly again in 2007, to 253. However, from this point in 2007 for the remainder of the decade, none of the movements in the mean Asian/Pacific Islander mean score for fourth graders for each Main NAEP represented a significant change over its predecessor. These movements were 253 (2007) to 255 (2009) to 256 (2011) to 258 (2013) to 257 (2015). All of these changes represented growth until the change from 2013 to 2015, at which point the change was a non-significant drop of 1 scale score point.

Performance gap analysis

In 2005 the gap score between White and Black fourth graders, as suggested by the difference in their NAEP scale score performance mean, was 26 score points, favoring White students. Across the decade, as both groups' scores grew, the difference remained at 25 or 26 scale score points, with two exceptions. In 2015, Black students had a non-significant gap gain of 1 point, but the 2015 score showed a significant gap loss of 2 points for White students, while Black students held steady. At the end of the decade the gap size was 24 scale score points.

The gap score between White and Hispanic fourth-grade students shrank 2 points between 2005 and 2015. At the opening of the decade, the gap was 20 points, a figure that continually shrank over the 10-year period to 18 points. The shifts were significant for 2007, 2009, and 2011, and then in 2013 the gap widened to 19 points for two years, until 2015, when White students lost 2 points while the Hispanic students lost 1. So, at the end of the decade, the score gap was 18 points, a reduction from 20 points in 2005.

The gap between White and Asian/Pacific Islander fourth-grade students went in a different direction in the decade from 2005 to 2015. In 2005, the difference between White students (246) and Asian/Pacific Islanders (251) was –5 points, with Asian/Pacific Islander students outperforming White students significantly. The gap widened to –7 points in 2011, then increased to –8 points in 2013, and finally ended the decade in 2015 at –9 score points. In the final transition from 2013 to 2015, mirroring what happened in the time period between the last assessment and the 2015 assessment, the Asian/Pacific Islander students' mean score decreased by 1 point, a non-significant change, while the White students' score decreased by a statistically significant 2 points, leading to the gap score of –9 points for White students at the end of the decade. This was a significant change from the opening of the decade, when the gap was –4 points, favoring the Asian/Pacific Islander students at the fourth-grade level.

Although one of the goals of the NCLB legislation was to eradicate differences among the various groups by 2015, the data show that this goal was not met. The change from 2000 to 2015 for the White-Black comparison was from 31 points to 24 points. The change for the White-Hispanic comparison was from 27 points to 18 points. The change for the White-Asian/Pacific Islander comparison from 2000 to 2015 was not computed because of the lack of a score for the latter group as a result of sample problems in 2000. From 2003 to 2015, the gap increased from –3 to –9, favoring the Asian/Pacific Islanders. Hence, the score gaps for the White-Black and White-Hispanic both closed, by 22.5% and 33.3%, respectively. At the same time, however, the gap between the White and Asian/Pacific Islanders widened by 80%.

NAEP 2015 Grade 8 Results

Figure 5 shows the national trend in eighth-grade student performance on the Main NAEP 2015 mathematics assessment. The eighth-grade average score in the 2015 assessment was 282 points. When the non-rounded value for 2015 is compared with the non-rounded value for the 2013 assessment, the difference is closer to 2 points than the 3 points indicated by the graph. A comparison of the exact values indicates that the two scores are statistically different ($p < 0.05$), with the difference indicating that the 2015 value is significantly lower than the level achieved by the eighth graders in 2013. As in the case of the grade 4 assessment, the educational meaning of this shift is another matter. Early analyses suggest that the content on NAEP remained true to the NAEP framework while content taught in the classroom rapidly moved toward the CCSSM objectives. Others will ascribe the decline to the increased focus on developing students' understanding of algorithmic work instead of simply ensuring their ability to execute calculations without emphasizing the need to understand why what they are doing makes sense. Yet others will note that the years when some topics are introduced in the curriculum have changed relative to the test in the time between 2013 and 2015. What is clear is that eighth-grade students in 2015 scored significantly lower in actual unrounded scores than students scored in 2011 and 2013, the same as students scored in 2009, and significantly higher than students scored in 2007. Analyses carried out at the American Institutes for Research indicate that the drop in performance at grade 8 can be explained in part by the first option given above and in part by other causes put forward (Daro, Hughes, and

Stancavage 2015; Hughes et al. 2013; Kane 2015). It appears that this difference can be explained by the research cited about the discrepancies in opportunity to learn some of the tested material in classrooms following rapid implementation of the CCSSM curricular plan with its differences in comparison with the NAGB Framework for the 2015 Main NAEP Mathematics Assessment.

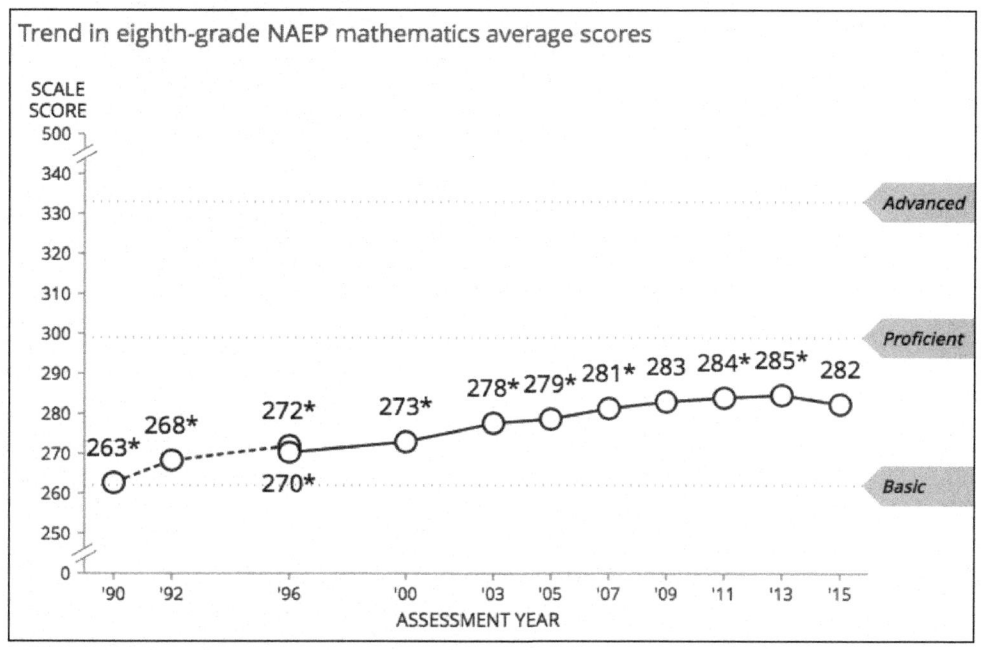

Fig. 5. Trend in eighth-grade NAEP mathematics mean scale scores. Graphic from the National Center for Education Statistics, *The Nation's Report Card: A First Look—Mathematics and Reading 2015*. Washington, D.C.: U.S. Department of Education, National Center for Education Statistics, 2015d. Used with permission.

Achievement levels

In 2015, 29% of eight graders performed at the Below Basic level, and 38% of eighth-grade students performed at Basic level. That is to say, in 2015, 67% of the nation's eighth-grade students were performing below the Proficient level. This result was 3% lower than the percentage of students performing below the eighth-grade Proficient level in 2013. The remaining 33% of students were performing at or above the Proficient level. This group splits into 25% at the Proficient level and 8% at the Advanced level. Performance at or above the Proficient level on NAEP assessments demonstrates that students are solid, both conceptually and procedurally. The percentage of eighth-grade students performing at or above the Proficient level in 2015 was 3% higher as compared with the percentage in 2005. Thus, although the results give signs of a slow gain, the gain is based on a slow shifting of students up from the Below Basic interval and into the Basic level, with a similar matching shift of a smaller percentage of students up from the Proficient level and into the Advanced level. However, in the end, only 33% of eighth graders had reached the level of at or above Proficient in 2015. Figure 6 shows the percentages of eighth graders at the various levels of mathematics achievement on NAEP assessments between 1990 and 2015.

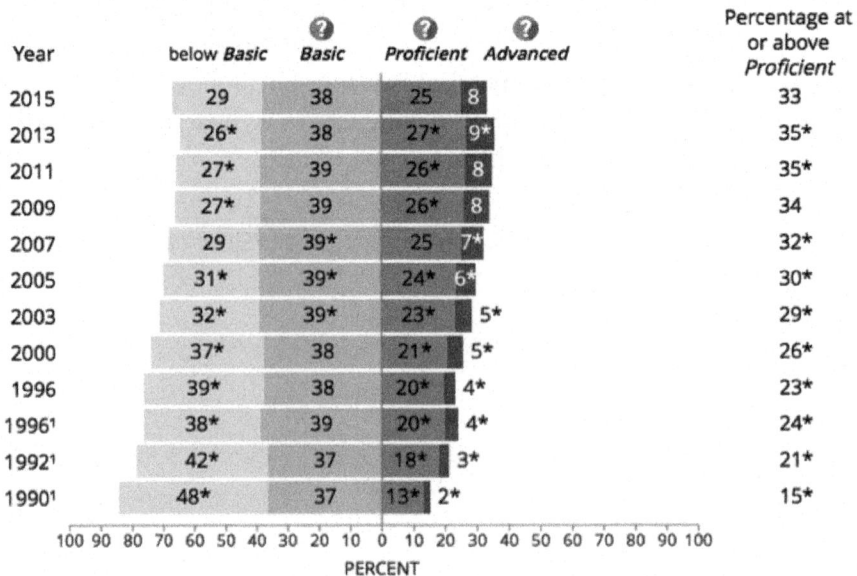

Year	below *Basic*	*Basic*	*Proficient*	*Advanced*	Percentage at or above *Proficient*
2015	29	38	25	8	33
2013	26*	38	27*	9*	35*
2011	27*	39	26*	8	35*
2009	27*	39	26*	8	34
2007	29	39*	25	7*	32*
2005	31*	39*	24*	6*	30*
2003	32*	39*	23*	5*	29*
2000	37*	38	21*	5*	26*
1996	39*	38	20*	4*	23*
1996[1]	38*	39	20*	4*	24*
1992[1]	42*	37	18*	3*	21*
1990[1]	48*	37	13*	2*	15*

Fig. 6. Percentage of grade 8 students at each NAEP achievement level by assessment year. Graphic from the National Center for Education Statistics, *The Nation's Report Card: A First Look—Mathematics and Reading Assessments 2015*. Washington, D.C.: U.S. Department of Education, National Center for Education Statistics, 2015d. Used with permission.

Gender comparisons

Comparisons for male and female eighth-grade performances in 2015 showed both male and female eighth graders at a mean performance level of 282. There was no significant difference in male and female performance in mathematics at the grade-eight level on the 2015 NAEP Mathematics Assessment. An examination of data from the years 2005–2015 shows that the performance differences in the groups are slowly growing smaller.

Race and ethnicity differences

As noted previously, comparisons of racial and ethnic differences are loaded with a number of correlated factors. Given the inability to cover them in a completely scholarly way, we present the nature of the trends in each group over time and leave the discussion of the gaps between different groups to other publications. Data exist that provide a basis to talk about trends across the past decade, 2005–2015, for students of White, Black, Hispanic, and Asian/Pacific Islander descent. The groups' percentages of the 2015 student eighth-grade population in 2015 were 52%, 15%, 24%, and 6%, respectively. The remaining groups, for which full information was not available, were Asians (5%), American Indians/Alaska Natives (1%), and students classified as being of two or more races (2%).

White student performance began the decade with a mathematics achievement mean scale score of 289 for 2005, and that score grew significantly, to 291 ($p < 0.05$), in 2007. It grew significantly again, to 293, in 2009, and remained unchanged at 293 in 2011. The mean scale score increased to 294 in 2011, but the change was not significant statistically. White eighth-grade student performance dropped 2 points in 2015, to 292, a statistically significant decrease that contributed heavily to the national decline, owing to the large percentage of the population belonging to this group. Further, it was at a level not significantly different from the 291 achieved in 2007, almost a decade earlier.

Black eighth graders started the decade in 2005 at 255, a significant increase from the 2003 assessment, and significantly increased their performance again in 2007, to 260.

In 2009, Black eighth graders experienced a slight, but significant, increase, to 261. Then they continued to make slight, but insignificant, increases, moving from 261 to 262 from 2009 to 2011 and from 262 to 263 in 2013. However, in 2015, the level of Black performance returned to 260, a significant drop from the 2013 result. In fact, this was a return to the 2011 achievement score level for Black eighth graders. This slippage in 2015 contributed to the overall decline in the national performance in eighth-grade mathematics achievement.

Hispanic eighth graders began the decade at 262, which represented a significant increase ($p < 0.05$) over their 2003 performance, and their mean score increased significantly again in 2007, to 265. This increase was followed by two more significant increases from 2007 to 2009 (266) and from 2009 to 2011 (270). Then in 2013, the NAEP assessment showed growth from the 270 to 272, but this increase was statistically insignificant. In 2015, the change was downward, from the 272 in 2013 to 270, a movement that was statistically significant. This decrease to 270 was a return for Hispanic eighth graders to their 2011 achievement score level. Given the sizeable proportion of the eighth-grade population that is Hispanic, this difference accounted for a sizeable part of the decrease in overall U.S. eighth-grade performance in 2015.

Asian/Pacific Islander students started the decade in 2005 at 295, a significant increase from their 2003 performance, and then these students moved to 297 in 2007, an insignificant increase. Their 2009 performance of 301 was a statistically significant increase from the 2007 level. Assessment results for this population from 2011 showed an insignificant increase to 303 from 2009, and results from 2013 showed an insignificant increase to 306 from the score level in 2011. The Asian/Pacific Islander eighth graders' performance in 2015 stayed at the same level as in 2013, making the Asian/Pacific Islanders the only racial group or ethnicity that did not see a significant decrease in mathematics performance in 2015.

Achievement gap analysis

The comparison of gaps between the performance of two groups across time is a measure, but only one, of the direction of movement to equality or inequality in an educational system. The target of achieving gap equality through a steady decrease in gaps across the board was a major goal of the NCLB legislation. In 2005 the gap score between White and Black eighth graders, as suggested by the difference in their NAEP scale score performance mean, was 34 eighth-grade NAEP scale score points, favoring the White students. Across the decade, as both groups' yearly scores slowly increased, their gap score (White score minus Black score) decreased to a 31-point difference in 2011 and held there in 2013. The 2015 gap score increased, but not significantly, to a 32-point difference. Thus, at the end of the decade in 2015, the size of the gap between White and Black students was 32 scale score points, a statistically significant improvement from 2005.

The gap score between White and Hispanic eighth-grade students shrank 7 points between 2005 and 2015. At the opening of the decade, the gap was 27 points, a figure that continually shrank over the 10-year period to 22 points. The shifts were steady from 2007 (26) to 2009 (26) to the level in 2011 (23). White-Hispanic gap scores moved one point closer in 2013 (22), a level that was matched by the 2015 gap score, 22. The last three gap scores in the sequence 2011, 2013, and 2015, were not significantly different, but all were significantly smaller than those viewed in 2005, 2007, and 2009. Hence, even after hitting a plateau at the end of the decade, the White-Hispanic eighth-grade gap score for Main NAEP mathematics did decrease over the decade from 2005 to 2015. Despite this movement in the right direction, a concern still remains about the lack of progress in movement in the last two NAEP assessment periods in that decade.

The gap between White and Asian/Pacific Islander eighth-grade students was similar to that indicated in the score analysis for White students and Asian/Pacific Islanders students at the fourth-grade level. Across the decade from 2005 to 2015, the Asian/Pacific Islander eighth-grade students consistently scored significantly higher in mathematics than White students for each individual Main NAEP assessment. In 2005, the difference was significant between White students (289) and Asian/Pacific Islander students (295) at –7 points when the exact values rather than rounded values were compared. The gap narrowed to –6 point in 2007, but widened to a –8 points in 2009 and continued widening across the remainder of the decade, from –9 in 2011, to –12 in 2013, and to –14 in 2015. In the final transition from 2013 to 2015, reflecting what happened nationally in the time period for all students, the Asian/Pacific Islander students' score held steady at 309, while the White students' score decreased 2 points. This accounted for the final widening of the gap score in the same transition.

Remembering that one of the goals of the NCLB legislation was to eradicate differences among groups by 2015, we note that the data show that the goal was not met. The change from 2000 to 2015 for the White-Black comparison was from 34 points to 32 points (a 5.9% decrease). The change for the White-Hispanic comparison was from 27 points to 22 points (a 18.5% decrease). The change for the White-Asian/Pacific Islander comparison from 2005 to 2015 was from –7 to –14, respectively—a 50% increase in the difference (gap) over the period, favoring Asian/Pacific Islanders. Hence, the score gaps between White and Black students and between White students and Hispanic students both decreased, while the gap between White students and Asian/Pacific Islander students increased.

NAEP 2015 Grade 12 Results

From a trend standpoint, the data for students in grade 12 from the 2009 NAEP mathematics assessment have to be treated as preliminary or initial. The results from the 2009 assessment at grade 12 showed a mean scale score of 153. This was 3 points higher than the corresponding score in 2005, and the difference was judged to be statistically significant (NCES 2009a). The 2013 grade 12 assessments consisted of a Main NAEP study at the grade 12 level but only a partial State NAEP study, which was open to states on a voluntary basis, given the limited nature of the state sample in the assessment in 2013. Thirteen states volunteered for the State NAEP portion of the NAEP 2013 grade 12 assessment of mathematics. Eleven of these states had also volunteered in 2009 for the last trial run of the new grade 12 NAEP assessment program, thus enabling an examination of the mathematics achievement differences for these states over the previous four-year period. The state samples were considered to form a "national" data set, and analyses were run to establish the representativeness of the "national" and pilot state samples as a basis for reporting. The analysis of the data was seen to support the idea that both met the individual-level criterion for "national" and for "13-state" reporting. The one exception was that the response rate for private schools did not meet the participation rate levels required by the sampling design. Thus, data from the grade 12 "nation" and participating "states" are reported only for outcomes for public schools for grade 12 in 2013.

The analysis of the national findings in 2013 in the National Grade 12 NAEP mathematics assessment showed that the mean score for grade 12 performance (153) represented a significant increase from the first administration in 2005 (150) but no increase from the score in the 2009 administration of the grade 12 test (153). The 2013 mathematics assessment scores within the varied racial and ethnic groups were all stable, with no significant change from their score levels achieved in the 2009 administration within groups.

Further, students who took more advanced mathematics courses scored higher than those who took lower-level mathematics courses or left the mathematics curriculum earlier in their high school studies. The mean achievement scores by course level were calculus (187), precalculus (165), algebra 2/trigonometry (143), geometry (125), and algebra 1 (112). Further, one must take into account the differential achievement rates for students who discontinue their formal education on or shortly after reaching the age of compulsory education but before graduating from secondary school (NCES 2011).

The 2013 NAEP grade 12 mathematics assessment was intended as a validation study of to determine whether the NAEP grade 12 assessment met the criterion for NAGB's two goals of being able to describe 12th graders' preparation for placement into (1) entry-level college credit courses without remedial coursework or (2) job-training programs without remedial coursework in mathematics or reading (Fields 2014). On the basis of the results of the 2013 grade 12 NAEP mathematics assessment, NAGB approved cut points on the NAEP scale for reading (302) and mathematics (163) that the validity study indicated would make them academically prepared for college.

NAGB believes that those these cut points appear to be supported by their validity study but will continue to study of the relationship between NAEP scores and the points at which students can be deemed ready to enter college without the need for remedial coursework or enter job-training programs without remedial coursework (Fields 2014). In establishing these score cut points, NAGB has indicated that the Main NAEP grade 12 mathematics assessment would continue, with regular alignments of its framework, to serve as a basis for the mathematics underlying the two goals set for using NAEP as a basis for measuring progress toward these goals while at the same time continuing a trend line for student performance in mathematics in the final year of secondary education (Fields 2014; NCES 2013d). The full results of the 2015 NAEP grade 12 mathematics assessment will be available in 2016.

NAEP 2015 Trial Urban District Assessment (TUDA)

Starting in 2003 and continuing in 2005, 2007, 2009, 2011, 2013, and 2015, the National Assessment has conducted a Trial Urban District Assessment (TUDA), involving some of the largest school districts in the nation. In light of the fact that one-quarter of the nation's youth live in urban areas, the success of urban students in preparing for postsecondary study in STEM fields is critical to the quality of the nation's workforce in coming decades. Hence, special monitoring of mathematical learning opportunities for these youth is crucial to ensure their progress as well as continued U.S. economic progress (NCES 2013b).

The results of TUDA 2015 were reassuring in that they showed significant progress in some of the largest districts that had been involved in TUDA assessments and had implemented innovative programs directed toward the goals of STEM program improvement. Table 8—shown in two parts, (a) and (b)—contains trend data in mean mathematics scores for grades 4 and 8 from the 2015 TUDA assessment for the nation, large cities, and twenty-one participating urban districts. The TUDA 2015 results for 2015 are included in *The Nation's Report Card: A First Look: Mathematics and Reading 2015* under the heading for districts (NCES 2015d).

Table 8
*Trend in mean NAEP mathematics scores for TUDA districts: Grade 4 and grade 8,
2005–2015*

a. Grade 4

	2005	2007	2009	2011	2013	2015	% ≥ Prof 2015
Nation (public)	237	239	239	240	241	240*	39*
Large cities	228	230	231	233	235	234	32
Albuquerque	dnp	dnp	dnp	235	235	231	28
Atlanta	221	224	225	228	233	228	26
Austin	242	241	240	245	245	246*	47*
Baltimore City	dnp	dnp	222	226	223	215	12
Boston	229	233	236	237	237	236	33
Charlotte	244	244	245	247	247	248*	51*
Chicago	216	220	222	224	231	232	30
Cleveland	220	215	213	216	216	219	13
Dallas	dnp	dnp	dnp	233	234	238*	24
Detroit	dnp	dnp	200	203	204	205	5
District of Columbia (DCPS)	211	214	220	222	229	232	33
Duval County (FL)	dnp	dnp	dnp	dnp	dnp	243*	41*
Fresno	dnp	dnp	219	218	220	218	14
Hillsborough County (FL)	dnp	dnp	dnp	243	243	244*	43*
Houston	233	234	236	237	236	239*	36
Jefferson County (KY)	dnp	dnp	233	236	234	236	34
Los Angeles	220	221	222	223	228	224	22
Miami-Dade	dnp	dnp	236	236	237	242*	41*
New York City	231	236	237	234	236	231	26
Philadelphia	dnp	dnp	222	225	223	217	15
San Diego	232	234	236	239	241	233	31

b. Grade 8

	2005	2007	2009	2011	2013	2015	% ≥ Prof 2015**
Nation (public)	278	280	282	283	284	281*	32*
Large cities	265	269	271	274	276	274	26
Albuquerque	dnp	dnp	dnp	275	274	271	21
Atlanta	245	256	259	266	267	266*	20
Austin	281	283	287	287	285	284*	35*
Baltimore City	dnp	dnp	257	261	260	255*	12
Boston	270	276	279	282	283	281*	34*
Charlotte	281	283	283	285	289	286*	39*

b. Grade 8, *continued*

	2005	2007	2009	2011	2013	2015	% ≥ Prof 2015**
Chicago	258	260	264	270	269	275	25
Cleveland	249	257	256	256	253	254*	9
Dallas	dnp	dnp	dnp	256	253	254	20
Detroit	dnp	dnp	238	246	240	244*	4
District of Columbia (DCPS)	245	248	251	255	260	258	17
Duval County (FL)	dnp	dnp	dnp	dnp	dnp	275	22
Fresno	dnp	dnp	258	256	260	257*	12
Hillsborough County (FL)	dnp	dnp	dnp	282	284	276	27
Houston	267	273	277	279	280	276	27
Jefferson County (KY)	dnp	dnp	271	274	273	272	26
Los Angeles	250	257	258	261	264	263*	15
Miami-Dade	dnp	dnp	273	272	274	274	26
New York City	267	270	273	272	274	275	27
Philadelphia	dnp	dnp	265	265	266	267*	20
San Diego	270	272	280	278	277	280*	32*

Notes:

*2015 score significantly larger from U.S. large city mean in 2015 (grade 4 = 234, grade 8 = 274).

** % ≥ "Prof" is percentage of students in public schools at or above Proficient in 2015.

"Large cities" includes students from all cities in the nation with populations of 250,000 or more, including the participating TUDA districts.

(Data from National Center for Education Statistics, *Trial Urban District Assessment Snapshot Reports,* Washington, D.C.: NCES, 2015).

In 2015, TUDA results for grade 4 students in mathematics in seven participating urban districts were significantly higher than those from large cities nationwide. The performances of grade 8 urban youth from five TUDA cities had higher means than their peers on the NAEP math assessment in other TUDA locations in 2015 (NCES 2015) These results are encouraging but indicate additional ground still needs to be gained to achieve the necessary STEM levels in urban settings.

NAEP 2012 Long-Term Trend Assessment Study

Although the national NAEP assessments and their frameworks are designed to change as the curriculum and school programs change, the National Center for Education Statistics also administers an additional NAEP assessment, the NAEP Long-Term Trend Assessment (LTTA), to a nationally representative sample of students. Initiated in 1973, this assessment used exactly the same test over time under the same conditions through 1999. Because the early NAEP assessments drew samples of 9-, 13-, and 17-year-olds, the LTTA continues to collect data at these age levels, even though Main NAEP has moved to grade 4, grade 8, and grade 12 samples for ease in dealing with policy issues more directly.

The NAEP trend assessment provides valuable information on whether students' performance on items considered important in 1973 (such as paper-and-pencil computation skills, direct application of measurement formulas in geometric settings, and the use of mathematics in everyday situations involving time and money) has changed over time (NCES 2013e; Perie, Moran, and Lutkas 2005). Data from the NAEP long-term trend assessments in mathematics from 1973 to 2012 are shown in figure 7.

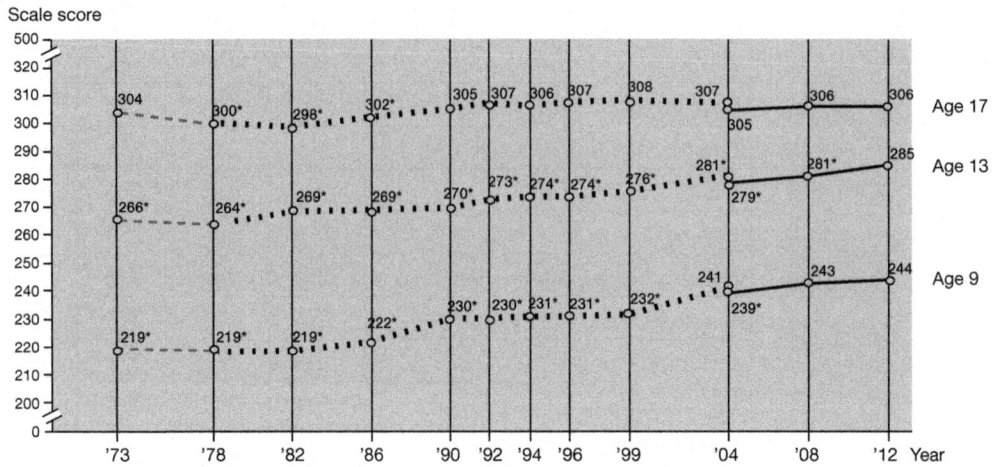

Fig. 7. Trend in NAEP LTTA mean mathematics scores for 9-, 13-, and 17-year-old students. Data and graph from National Center for Education Statistics, *The National Report Card: Trends in Academic Progress 2012*, Washington, D.C.: U.S. Department of Education, National Center for Education Statistics, 2013e. Graph used by permission.

LTTA analyses over time have to be considered carefully, given the changes that were implemented with the 2004 assessment. Two statistical bridge studies were conducted, and the comparable content between the two has allowed the extension of the age-based trend lines as shown in figure 7, with the solid shading to represent comparable levels of performance. A dotted line reflects data for the period from 1978 to the changes that went into effect with the 2004 LTTA administration, and the solid line represents data from 2004 to the 2012 assessment. A comparison from 1973 to 1978 was made, but too many factors changed in that period to incorporate it as part of the first trend line. Thus, it is possible to make statistically meaningful comparisons over the time from 1978 to the present, but one must be careful from a validity standpoint when making comparisons that involve the content areas where changes were made.

In 2004, 2008, and 2012, the LTTA for mathematics was conducted by using the new long-term trend assessment test forms. The new assessment can change gradually over time, like Main NAEP, contrary to the invariant assessment that was used from 1973 through 1999. A bridge assessment has indicated that the continuation of the trend line between the old and new assessments is appropriate (Perie, Moran, and Lutkas 2005). Figure 7 contains the data for the performances of students at each of the three age ranges in the NAEP LTTA. The 2008 level of performance for both 9- and 13-year-old groups is statistically higher than that of the same age groups at every testing period from 1999 or earlier. The trend line for 17-year-olds shows a pattern of insignificant variation from 1990 to 2008 (Rampey, Dion, and Donahue 2009). These findings indicate that, on average, elementary and middle school students in 2008 had a better command of the fundamental concepts and skills deemed important in 1978 than their age-related peers across the history of the assessment. The 17-year-old group, with the slight exception observed in the 1978–1986 period, showed no appreciable growth or decline in their command of these basic concepts and skills over the period constituting the history of the assessment.

The LTTA was repeated again in 2012. The data from that cycle provides additional support for the interpretations of the trends. For a full analysis, one needs to read the reports of the 2004, 2008, and 2012 LTTA studies in mathematics (Perie, Moran, and Lutkus 2005; Rampey, Dion, and Donahue 2009; NCES 2013e). The three data points for age-group performance provide a third data point for each of the new trend lines. At the age 9 level, the 2012 performance scale score was judged to be significantly higher than the one for 2004 but not significantly different from 2008. The short trend line shows a modest increase over the eight-year span from 2004 to 2012. At the age 13 level, the 2012 performance scale score was judged to be significantly higher than the 2004 and 2008 scale score levels. This indicates a slightly stronger trend of improvement since 2004. At the grade 12 level, there were no significant differences in scale score performance from 2012 over performance for grade 12 students in either 2004 or 2008.

When one looks back to the trend lines from the 1978 assessment to the 2012 assessment, one sees the long-term changes in student performance in mathematics over a relatively constant test. At the age 9 level, the 2012 data point was judged significantly higher than any of the other data points for that age level for each assessment from 2004 back to 1978. There was no significant difference between the 2012 and 2008 assessment. This is a strong indication that U.S. students at age 9 perform, on average, better than their parents and grandparents did on the "basics" in mathematics.

At the age 13 level, the evidence is even stronger. Under the metrics for both trend lines, the mean score scale value for 2012 for age 13 students was judged significantly higher than the mean performance for all of the assessments from 2008 back to 1978. This again affirms that in 2012 13-year-old U.S. students, on average, performed better on the "basics" than their counterparts in their parents' and grandparents' generations did.

The LTTA results provide a different profile for U.S. students at age 17. At this level, the outcomes are much the same as those seen in the grade 12 Main NAEP assessments, and they paint a picture of little change over time. Here, the short LTTA trend line shows no difference in mean scale scores for 2004, 2008, and 2012.

The grade 12 results aside, the NAEP LTTA results confirm that at the age 9 and age 13 levels, today's U.S. schoolchildren are performing at a significantly higher level than their parents and grandparents did in mathematics. This speaks positively to progress in curricula and teaching at these levels. The next LTTA is scheduled for 2024.

State NAEP 2012 Results

Table 9 illustrates the vast differences among states' mean achievement scores and the percentages of students reaching the level identified as Proficient.

A comparison of the 2015 state-level data with 2013 state-level NAEP scores for grade 4 found that students in Mississippi, the Department of Defense schools, and the District of Columbia had significantly higher scores than they had in 2013. Meanwhile, fourth graders in the states of Montana, Connecticut, Rhode Island, New York, Georgia, Maine, New Hampshire, Minnesota, Delaware, Vermont, Arkansas, Kansas, Colorado, Hawaii, and Maryland were found to have significantly lower scores in 2015 than they had in 2013. Although the difference scores are shown as the differences of their rounded values, the state difference scores are calculated on the basis of their unrounded difference scores and the standard errors of the respective scores (NCES 2015d).

Table 9

Mean State Main NAEP scores and percentage Proficient for 2015

Jurisdiction	GRADE 4		GRADE 8	
	Mean NAEP scale score—2015	Percent Proficient or above—2015	Mean NAEP scale score—2015	Percent Proficient or above—2015
NATION	240	40	282	33
NATION (public)	240	39	281	32
Alabama	231	26	267	17
Alaska	236	35	280	32
Arizona	238	38	283	35
Arkansas	235	32	275	25
California	232	29	275	27
Colorado	242	43	286	37
Connecticut	240	41	284	36
Delaware	239	37	280	30
Florida	243	42	275	26
Georgia	236	35	279	28
Hawaii	238	38	279	30
Idaho	239	38	284	34
Illinois	237	37	282	32
Indiana	248	50	287	39
Iowa	243	44	286	37
Kansas	241	41	284	33
Kentucky	242	40	278	28
Louisiana	234	30	268	18
Maine	242	41	285	35
Maryland	239	40	283	35
Massachusetts	251	54	297	51
Michigan	236	34	278	29
Minnesota	250	53	294	48
Mississippi	234	30	271	22
Missouri	239	38	281	31
Montana	241	41	287	39
Nebraska	244	46	286	38
Nevada	234	32	275	26

Table 9. *Continued*
Mean State Main NAEP scores and percentage Proficient for 2015

Jurisdiction	GRADE 4		GRADE 8	
	Mean NAEP scale score—2015	Percent Proficient or above—2015	Mean NAEP scale score—2015	Percent Proficient or above—2015
New Hampshire	249	51	294	46
New Jersey	245	47	293	46
New Mexico	231	27	271	21
New York	237	35	280	31
North Carolina	244	44	281	33
North Dakota	245	45	288	39
Ohio	244	45	285	35
Oklahoma	240	37	275	23
Oregon	238	37	283	34
Pennsylvania	243	45	284	36
Rhode Island	238	37	281	32
South Carolina	237	36	276	26
South Dakota	240	40	285	34
Tennessee	241	40	278	29
Texas	244	44	284	32
Utah	243	44	286	38
Vermont	243	43	290	42
Virginia	247	47	288	38
Washington	245	47	287	39
West Virginia	235	33	271	21
Wisconsin	243	45	289	41
Wyoming	247	48	287	35
Other jurisdictions				
District of Columbia	231	14	263	19
DoDEA	248	37	291	40

Comparison of state-level grade 8 NAEP 2015 data with same-state data from 2013 showed that in no state in 2015 did students score significantly higher than students in that same state had scored in 2013. Comparison of state-level grade 8 NAEP 2015 data with corresponding data from the same state from 2013 showed that students in 22 states scored significantly lower in 2015 than students in that same state had scored in 2013. These 22 states were North Dakota, Delaware, South Dakota, Rhode Island, Idaho, Kentucky, West Virginia, Nevada, Washington, Maryland, Massachusetts, Maine, South Carolina, Colorado, North Carolina, Louisiana, Texas, Ohio, Vermont, Florida, Kansas, and Pennsylvania. The latter 12 states had rounded declines of 4 points (NCES 2015d).

There were no comparable state-level comparisons for grade 12. Data for some states at grade 12 will be available as part of the 2015 Grade 12 NAEP release in 2016.

Consortia for the Assessment of Common Core–Related Achievement

As mentioned in Chapter 2, two assessment consortia of states were formed as a result of a competition that was part of the Race to the Top program of the Department of Education. Although linked to the Department of Education for base funding, these consortia are now drawing significant fees from the states that belong to them as the assessments go online. Further, although the two consortia are not directly or fiscally linked to the Common Core State Standards program, the tests developed by the consortia are tied to the assessment of the CCSSM program.

Partnership for Assessment of Readiness for College and Careers (PARCC)

The first full-scale administration of the PARCC had 5 million students taking their tests across grades 3–8 and in grade 11 during the 2014–15 school year. Eleven member states and the District of Columbia had students taking part in that school year's testing. In addition, New York City had a pilot program of 5,000 students from 25 of its schools taking the PARCC tests.

The PARCC assessments are delivered at two different points in the school year. In early spring, the examination calls on students to work problems that take more time than conventional test items, such as multi-step mathematics problems that require students to draw on knowledge about a given applied setting. At the same time, the PARCC assessments also include problems that assess students' capabilities to call on and apply their procedural prowess in mathematics. Near the end of the school year, students complete another assessment, which consists of short constructed-response items and multiple-choice items. These items focus on ensuring that students have mastered the key concepts and procedures and can demonstrate their understanding in short applications, whereas the earlier test focuses on reasoning, sequencing information, and applying strategies while solving assessment items that are more context based (PARCC 2015b).

Smarter Balanced Assessment Consortium (SBAC)

The first full-scale administration of the SBAC assessment had more than 6 million students taking tests across grades 3–8 and in grade 11 during the spring summative assessments in the 2014–15 school year. Students from 18 member states and selected from schools from the Bureau of Indian Education and the U.S. Virgin Islands took part in this testing.

Like the PARCC assessments, the SBAC assessments are delivered at two different points in the school year. The consortium has developed a digital library of items and rubrics for teachers to use during the school year as they are teaching associated CCSSM content and mathematical practices. The digital library also includes lesson-planning ideas to illustrate teaching practices to support students in developing skill in using the CCSSM mathematical practices.

The assessment package also contains optional interim assessments. These assessments are computer delivered and contain performance items based in complex real-world settings that call on students to apply their understanding of mathematical concepts and their procedural knowledge, along with problem-solving strategies. The optional assessments are tied to the CCSSM curriculum, and teachers can decide when it is appropriate to employ them. They contain scoring rubrics and links to help teachers understand how their students' performance compares with what might be expected on the summative assessment.

The summative assessment is administered in the final 12 weeks of the school year. This test has two parts. The first portion is a computer-delivered adaptive test that selects problems matching the students' response patterns to accurately determine what they know and where they are still developing command of the curricular contents. The second portion of the summative test consists of performance items (conceptual and procedural) that are also computer delivered but not adaptive in nature (SBAC 2015).

College Entrance Examinations

Typically, a student in the United States applies for college in the twelfth grade, the last year of high school. The selectivity of colleges in the United States varies, from community colleges and postsecondary institutions that require no more than a high school diploma or its equivalent to selective colleges that may accept 10% or fewer of their applicants. The selectivity of an institution sometimes varies with the academic major that a student expresses an intent to pursue. Because college entrance examination scores provide the only easily quantifiable and comparable measure for students coming from different high schools and different areas of the country, they are often given great importance by colleges. As a result, most college-intending students in the United States take a college entrance examination during their junior or senior year of secondary school.

Two such major and independent college admission examinations exist in the United States. The SAT test, administered by the College Board, is more common in the east, south, and west. Institutions in the middle portion of the country more commonly prefer the ACT test, administered by ACT, Inc. Forty-nine percent of graduating seniors took the SAT in the 2012–13 school year, and 59% of graduating seniors took the ACT in the same school year (Snyder and Dillow 2015; ACT 2015). The data on the percentages taking the tests have been influenced strongly in recent years by some states requiring that all their students take the ACT or SAT at least once during the eleventh or twelfth grade as part of their state assessment program (ACT 2015; College Board 2015e).

The SAT assesses high school students' general capabilities in critical reading and mathematics. A student's results for each of the two sections of the SAT are reported on a 200–800 scale. A student's overall reasoning ability is reported by the sum of the individual scale scores on a 400–1600 scale. The mathematics test employs multiple-choice items and open-ended items for which the response is gridded into a special array of bubbles on an optically scanned answer sheet. The mathematics portion of the test covers number and operations; algebra and functions; geometry; and statistics, probability, and data analysis. Beginning with the March 2005 administrations of the SAT, the coverage of the mathematics test was extended to include more items testing the content of second-year algebra and more advanced topics from geometry. However, the focus in these items, as in the previous versions, remains on students' critical reasoning skills (College Board 2011; Kobrin and Schmidt 2007).

The ACT assesses high school students' general subject-matter knowledge and college or workforce readiness in four skill areas: English, mathematics, reading, and science. The test is composed entirely of multiple-choice items, and each of the four skill areas is reported on a 1–36 scale. A general summary score on the same 1–36 scale is used to report a student's overall skill level (ACT 2015). Graduating seniors' mean mathematics performance on both the SAT and ACT has shown substantial improvement since 1995 (see table 10). The asterisk indicates scores on the new form of the SAT, beginning with 2006 data. These scores result from a bridge study that provides the basis for them to be reported on the same scale in a valid and reliable fashion for interpretation and comparison.

Table 10
National mean grade 12 scores for mathematics on the SAT and ACT exams

Year	Test	
	SAT-Math	ACT-Math
2000	514	20.7
2001	514	20.7
2002	516	20.6
2003	519	20.6
2004	518	20.7
2005	520	20.7
2006*	518	20.8
2007	515	21.0
2008	515	21.0
2009	515	21.0
2010	516	21.0
2011	514	21.1
2012	514	21.1
2013	514	20.9
2014	513	20.9
2015	511	21.0

*SAT content modified in 2006 and following years, but all data in the table are reported using scores adjusted to 2015 scale score values. The grade cohort for the senior year is seniors who graduated in the year 2015, had taken the SAT at least once, and the score used for their report SAT score is the most recent score test score in their records.

With the March 2016 SAT, the content and format of the test will change. The new SAT will still focus on problem solving, but it will place greater emphasis on linear equations and systems. The problem solving will give additional attention to thinking with data. Finally, more items will involve equations that require manipulation on the way to an answer. The items will focus on mathematics content, including geometry and trigonometry, that are related to both college readiness and career readiness and advancement (College Board 2015e).

The general public has come to view these mean scores for the secondary school graduating class in a given year as a barometer of how well the education system is performing as a whole, despite the fact that the examinations were not designed for that purpose and have obvious shortcomings when used as a single measure of students' mathematics competence and overall quantitative literacy. Both ACT, Inc. and the College Board provide an extensive study each year of the students in that year's graduating class who took either or both the SAT and ACT exams in their senior year (College Board 2015f; ACT 2015).

SAT Subject Tests

In addition to the SAT examination itself, the College Board also administers subject-area tests to allow students to gauge their progress as well as to have evidence to offer to colleges of their suitability for admission or higher course placement. Two of these course-level tests are the Mathematics Level 1 and Mathematics Level 2 examinations. Mathematics Level 1 assesses students' knowledge of the first three years of a college preparatory curriculum—two years of algebra and one year of geometry. Mathematics Level 2 covers this same material at an advanced level, with the addition of the study of trigonometry through radian measure, Pythagorean identities, the laws of sines and cosines, and the double angle formulas, plus the elementary functions extensions to real and complex numbers, solutions of systems with matrices, and representations and properties of solutions of systems. These two tests are described in more detail in College Board publications found on the Web, along with sample examinations, at http://www.SATSubjectTests.org. Students are allowed to use most graphing calculators from a published list that includes clear information on the types of more advanced calculators that are not allowed.

An examination of student performance on these tests over time gives a picture of student performance on a pair of standardized tests covering common content over time. Table 11 provides a look at student performance on the SAT Mathematics Level I and SAT Mathematics Level II examinations for 2010 and 2015. The data illustrate the growth that took place from 2010 to 2015 for both tests. It is impossible to link back further because of some changes made in the calculator-use regulations as the examinations were changed from Math IC and Math IIC, which allowed the use of calculators. The newer tests have some items that demand that test takers use calculators in order to work the problems in a reasonable amount of time. At both Mathematics Level I and Mathematics Level II, one sees a movement into the upper three score intervals. The distribution of these percentages of scores within the intervals also plays a role in the difference of the performances for each test from 2010 to 2015.

Advanced Placement examinations

As is discussed further in Chapter 8, the College Board offers a set of Advanced Placement examinations for students who study advanced, college-level courses following a syllabus provided by the College Board. This syllabus is developed by a committee of subject-matter experts for the content of the course at the undergraduate level. Additional subject-matter experts vet the content on a regular basis for concurrence and pacing relative to courses at the freshman or sophomore level in college. Students study the material for a year under the leadership of a teacher who is knowledgeable about the syllabus and the manner in which students' understanding of the content of the course will be tested. The test is administered in April, near the end of the school year and is then graded by

Table 11

Distribution of percentages of student performance on the SAT Mathematics Level I and Mathematics Level II tests for the years 2010 and 2015

SAT Mathematics Level I			SAT Mathematics Level II		
Math Level I scores	2015	2010	Math Level II scores	2015	2010
750–800	9	6	750–800	38	25
700–749	19	16	700–740	15	13
650–699	18	17	650–690	16	15
600–649	17	20	600–640	14	15
550–599	13	15	550–590	10	14
500–549	11	11	500–540	5	9
450–499	7	7	450–490	2	6
400–449	4	5	400–440	1	2
350–399	2	3	350–390	0	0
300–349	1	1	300–340	0	0
250–299	0	0	250–290	0	0
200–249	0	0	200–240	0	0
Mean	619	605	Mean	690	649
Total tests	65,319	85,109	Total tests	144,772	163,713

Data from College Board SAT Mathematics Test Archives (http://research.collegeboard.org/programs/ap/data).

a select group of teachers and university faculty who use a common rubric for the open-ended items. The three examinations for the yearlong Advanced Placement courses in areas of mathematics are the following (Chapter 8 provides more information about the content covered in each course):

- Calculus AB (covers topics in differential and integral calculus that are roughly equivalent those in a first-semester college calculus course)

- Calculus BC (measures students' understanding of the concepts of calculus, their ability to apply these concepts, and their ability to make connections among graphical, numerical, analytical, and verbal representations of mathematics)

- Statistics (covers major concepts and tools for collecting, analyzing, and drawing conclusions from data that are the focus of a one-semester, introductory, non-calculus-based college course in statistics)

A sample of data for the 10 years from 2006 to 2015 appears in table 12, which illustrates student performance on the three examinations and gives the number of students taking each test in the years shown, the mean score for the students, the percentage of all AP tests taken that year, and the percentage increase in the number of students taking that course in the specified year over the previous year's total. The tests are taken by students from across all four years of high school, as well as some very accelerated special students. Although the scores are scaled, there is no evidence of a decrease in overall student performance on the examinations over the decade of results displayed. Students taking

Table 12

A decade of data describing performance on Advanced Placement Calculus AB, Calculus BC, and Statistics

Year	AP Calculus AB test				AP Calculus BC test				AP Statistics test			
	Tests in year	Mean score	Percent of total tests	Percent increase from prior year	Tests in year	Mean score	Percent of total tests	Percent increase from prior year	Tests in year	Mean score	Percent of total tests	Percent increase from prior year
2006	197,181	3.03	8.5%	6%	58,603	3.72	2.5%	8%	88,237	2.86	3.8%	15%
2007	211,693	2.94	8.4%	7%	64,311	3.71	2.5%	10%	98,033	2.80	3.9%	11%
2008	222,835	3.03	8.1%	5%	69,103	3.72	2.5%	7%	108,284	2.86	4.0%	10%
2009	230,588	2.99	7.9%	3%	72,965	3.72	2.5%	6%	116,876	2.83	4.0%	8%
2010	245,867	2.81	7.7%	7%	78,998	3.86	2.5%	8%	129,899	2.84	4.0%	11%
2011	255,357	2.82	7.4%	4%	85,194	3.77	2.5%	8%	142,910	2.82	4.1%	10%
2012	266,994	2.97	7.2%	5%	94,403	3.87	2.6%	11%	153,859	2.83	4.2%	8%
2013	282,814	2.96	7.2%	6%	104,483	3.73	2.7%	11%	169,508	2.80	4.3%	10%
2014	294,072	2.94	7.0%	4%	112,113	3.81	2.7%	7%	184,173	2.86	4.4%	9%
2015	302,532	2.86	6.8%	3%	118,707	3.72	2.7%	6%	195,526	2.80	4.4%	6%

From *AP Program Summary Report 2015* (College Board 2015b).

the examinations are sometime offered advanced placement in the calculus program at a university; in other cases, students are given credit based on their performances on the AP tests. Depending on the institution, these outcomes can range from no credit given but advanced placement in the second-semester course to credit given for the first two semesters of calculus.

There has been concern in recent years about the decline in scores on the mathematics portions of the SAT and the lack of improvement in ACT scores. The most prevalent explanation is that the numbers of students who are now attending college have increased, and they include many who would not have been traditional college attendees but are now taking the test and hoping to attend. At the same time, a large number of students taking the tests have no intention of attending college but are forced to take them as part of their state's graduation requirements. Both of these cases channel students with less mathematics preparation into the group of students sitting for the examinations, and their numbers also contribute to a lowering of the mean score.

Chapter 5: U.S. Participation in International Studies of Mathematics: Achievement and Context

International comparisons in mathematics education began in 1908 with a study commissioned by the International Commission on Mathematics Instruction to undertake a comparative study on the methods and plans for teaching mathematics in secondary schools. The study consumed six years and filled a long series of volumes with 310 reports from 18 countries (ICMI 2015). The field of international comparative studies of student achievement, with accompanying information on background factors, including within and between schools data, curricula, and teacher education programs, has come a long way since that beginning.

International Association for the Evaluation of Educational Achievement Studies

The first set of international comparisons of student achievement in mathematics that garnered public attention in the United States in recent memory was the First International Mathematics Study (FIMS), which collected data in 12 nations in 1964 from 13-year-olds and students in their final year of studies in secondary schools (Husén 1967). This study was conducted by the International Association for the Evaluation of Educational Achievement (IEA), a consortium of national ministries of education and principal educational evaluation institutions within member countries. The results of this study popularized the factor known as *opportunity to learn* (OTL) and highlighted it as a major causal factor for consideration in the study of student achievement. In the United States the study was given significant coverage in the *New York Times* as evidence of the failure of the New Math, confirming the lack of understanding of statistical comparisons and causal analysis among the general public in the United States (Hechinger 1967).

This study was followed in 1980–1982 by the Second International Mathematics Study (SIMS), which involved 20 countries and examined curricula, classroom practices, and student achievement of 13-year-olds and students who were in their final year of secondary school and were studying mathematics as a substantial part of their academic program of studies. The SIMS study examined the components—curricula, classroom practices, and student achievement—of mathematics education at three levels: intended, implemented, and achieved (McKnight et al. 1987). The results of this study at the international level suggested that IEA should establish a regular, cyclical set of studies with an international scope to analyze trends and examine changes in school mathematics, science, and reading curricula.

The Third International Mathematics and Science Study, conducted in 1995, was the first study launched with this model in mind and across the two disciplines of mathematics and science, three levels of schooling, and involving 46 countries and over half a million students in five grades. The acronym for the study (TIMSS) became the continuing acronym for the cyclical studies, now known and recognized as the Trends in International Mathematics and Science Study. The sequence of studies in mathematics began with the mathematics study in 1995 and continued with studies in 1999, 2003, 2007, 2011, and 2015 (IEA 2015). The reports of these studies usually followed in one or two years because of the time required for data cleaning and analysis involving a multitude of players (Beaton et al. 1996; Mullis et al. 1997, 1998; Mullis et al. 2004; Mullis et al. 2008a; Mullis et al. 2012).

Reports of these studies can be retrieved from the TIMSS study website, maintained at Boston College in Boston, Massachusetts (http://timssandpirls.bc.edu). A listing of additional publications at the website indicates the variety of side studies conducted in conjunction with these comparisons. One of the most useful study publications is the *TIMSS 2011 Encyclopedia* (Mullis et al. 2008b), which provides an in-depth view of the national contexts and educational structures for the participating nations and is a must-read for anyone who wishes to begin a serious study of the findings of the 2011 study (Mullis et al. 2012b). Also of interest is the most recent TIMSS assessment framework, which provides the design of content and cognitive processes underlying the program (Mullis et al. 2015). The TIMSS website also provides a location for downloading the analyses of the varied studies across time. The following discussion focuses on the TIMSS 2011 mathematics study—the latest analyzed study available at the time of writing. The TIMSS 2015 study results will appear and be available in spring 2016.

TIMSS 2011

The 2011 Trends in International Mathematics and Science Study (TIMSS) assessed students' mathematics achievement in grade 4 in 45 countries and at grade 8 in 38 countries. In addition to countries, a number of subnational jurisdictions—including the public school systems in several U.S. states (Alabama, California, Colorado, Connecticut, Florida, Indiana, Massachusetts, Minnesota, and North Carolina)—also participated in TIMSS as separate education systems. U.S. national results and those for the participating states can be found in the National Center of Educational Statistics TIMSS report of U.S. school mathematics performance in the study at grades 4 and 8 (Provasnik et al. 2013) or in the TIMSS international report (Mullis et al. 2012). The TIMSS assessments are curriculum based and measure what students have actually learned against the subject matter that is expected to be taught in the participating countries by the end of grades 4 and 8 (Snyder and Dillow 2015). Such assessments differ from assessments that focus on students' problem-solving skills or ability to reason in context alone. The TIMSS assessments consist of multiple-choice and short-answer items, as well as items requiring an extended constructed answer to the question or questions presented. The overall results on the achievement of a participating nation's students is presented as a mean score for the country on a 0- to 1000-point scale that has a mean performance level of 500 points and a standard deviation of 100 points. On the 2011 TIMSS, the average mathematics scores of U.S. fourth graders (541) and eighth graders (509) were higher than the scale score mean value of 500.

Fourth-grade performance

In 2011, U.S. fourth graders scored higher in mathematics, on average, than their counterparts in 37 countries and lower than those in 3 countries. Average mathematics scores in the other 4 countries were not measurably different from the U.S. average (541). Singapore (606), Korea (605) and Japan (585) were the 3 highest-scoring countries. An additional 4 countries whose overall performances were not statistically different from that of the U.S. at grade 4 were Finland (545), the Russian Federation (542), the Netherlands (540), and Denmark (537). The lowest-scoring country had a mean score of 248.

When all 57 of the geographical regions, states, and other identifiable regions that participated in the TIMSS 2011 mathematics sample are considered, U.S. fourth graders were statistically tied with 8 countries/regions with mean performance values statistically above that of the United States and 5 other countries/regions whose mean performance scores were not statistically different from those of the U.S. fourth graders. Thirty-seven other participating country/region mean performance scores were statistically lower than that of the U.S. students.

Analysis of the performance of the two U.S. states participating at the grade 4 level found that fourth graders in North Carolina (554) ranked significantly ($p < 0.05$) above those of fourth graders in the United States overall (541). The performance of fourth graders in Florida (545) was not statistically different from that of fourth graders in the U.S. national sample.

An examination of change in the U.S. fourth-grade performances against that of other participating countries in 2007 and 1995 found that U.S. fourth graders' mathematics performances had increased 23 score points ($p < 0.05$) since the 1995 TIMSS assessment and had increased 12 points ($p < 0.05$) since the 2007 assessment. Five educational systems had larger gains since the 2007 assessment and the 2011 assessment. However, all five of these were countries that scored significantly lower ($p < 0.05$) than the United States in 2007.

Analysis of the data on the content being assessed in the TIMSS grade 4 study by content subtest areas found that U.S. grade 4 students' performances ranked essentially the same as their overall test performance value of 541. Their performance means by content area were *number* (545), *geometric shapes and measures* (535), and *data display* (545). These values indicate that U.S. students performed significantly better than their overall average in the content areas of *number* and *data display* and significantly below in the content area of *geometric shapes and measures*. An examination of the trend in overall performance from 2007 to 2011 showed significant growth for the U.S. students in the content areas of *number* and *geometry* across the two assessments. Of the top 8 countries, 2 did not participate in 2007, so it was impossible to check their trend. However, only Japan in the group performing above the United States and the United States in its direct comparison group showed content trend areas with significant growth.

Analysis that considered the performance of all the geographical countries/regions, jurisdictions, or entities in the subtests *number* (543), *geometric shapes and measures* (535), and *data display* (545) found that the United States held essentially the same ranking position relative to other countries as it had in the earlier, country-only ranking. In *number*, the United States continued to perform below the top three countries, although its score, over time, had increased significantly. In *geometric shapes and measures*, it also ranked below Finland (543), which joined the top-performing group for this subtest. The Netherlands joined the top-performing group for the *data display* subtest (559) and thus was ahead of the United States in this subtest area. Overall, the U.S. performance on this three-assessment content subtest comparison showed no major weakness relative to the content subtest areas. Instead, the U.S. performance was at or significantly above the level of the majority of the nation's peers internationally.

At the fourth-grade level, the U.S. mean scale score for males was 9 points higher than that for females, making U.S. male performance at this level significantly greater statistically than that of U.S. females. At the state level of participation, fourth-grade males in both North Carolina and Florida, the two states participating at this level of TIMSS, significantly outperformed females by 12 points and 7 points, respectively.

Eighth-grade performance

In TIMSS 2011's assessment of mathematics performance at the eighth-grade level, the mean U.S. mathematics performance score was statistically above the mean scores of students in 27 countries in 2011 and statistically below that of students in 4 countries. U.S. eighth graders' mean performance on the eighth-grade PISA assessment in 2011 was 509. The four countries performing significantly higher than the U.S. mean performance were Korea (613), Singapore (611), Japan (570), and the Russian Federation (536). Six other countries' mean eighth-grade mathematics performance scores were not measurably different from the U.S. average. These countries were Israel (516), Finland (514), Hungary (505), Australia (505), Slovenia (505), and Lithuania (502).

The nine U.S. states participating as state-level jurisdictions had a more varied performance on the TIMSS score scale. Compared with the U.S. overall performance of 509, the states' performances break down as follows:

- Significantly higher than the U.S. mean performance: Massachusetts (561), Minnesota (545), North Carolina (537), and Indiana (522)

- Not significantly different from the U.S. mean performance: Colorado (518), Connecticut (518), and Florida (513)

- Significantly lower than the U.S. mean performance: California (493) and Alabama (466)

Analysis of change in the U.S. eighth-grade performances in 2011 against those of the other participating 34 countries that had also participated in both 2007 and 1995 found that U.S. eighth graders' mathematics performances was 1 score scale point above that of the performance of U.S. eighth graders on the 2007 assessment. This difference was not statistically significant. However, when the U.S. performance on the 2011 TIMSS was compared with the U.S. performance on the 1995 TIMSS assessment, a significant ($p < 0.05$) increase of 17 score scale points was observed.

Five educational systems had larger gains since the 2007 assessment and the subsequent assessment in 2011. Whereas the three countries ranking above the United States at grade 4 were the same in both 1995 and 2012, at the eighth-grade level, the Russian Federation moved from being not different from the United States in 2007 to performing at a significantly higher level than the United States in 2011, and at the same time, Australia and Slovenia moved from performing significantly below the United States in 2007 to performing at a level not different from the United States in 2011. Only two of the U.S. states had participated in at least one of the earlier TIMSS assessments. Massachusetts participated in the 2007 assessment, achieving a scale score of 547. In 2011, Massachusetts scored 561, indicating statistically significant growth between the two assessments. Furthermore, Massachusetts continued to perform significantly above the level of the United States mean score. Minnesota, with its scale score performance of 545 in 2011, also performed significantly higher than it had in both 1995 (518) and 2007 (532).

When the content assessed on the TIMSS grade 8 assessment is examined by content subtest areas, U.S. grade 8 students' performances are seen to rank essentially the same as their overall test performance value (509). Their performance means by content areas were *number* (514), *algebra* (512), *geometry* (485), and *data and chance* (527). These values indicate that U.S. students performed significantly better than their overall average in the content areas of *number, algebra, data and chance*. However, U.S. eighth graders performed significantly below their overall performance in the content area of *geometry*. Among the countries not significantly different from the United States overall, the U.S. eighth graders were the only ones performing significantly above their overall average in three or more areas. Table 13 presents the mean eighth-grade content scale scores in mathematics for states in TIMSS 2011.

Table 13
Mean grade 8 mathematics content scale scores for states in TIMSS 2011

State	Total 2011 TIMSS score	Number score	Algebra score	Geometry score	Data and chance score
TIMSS mean scale score	500	500	500	500	500
United States	509	514	512	485	527
Massachusetts	561	567	559	548	584
Minnesota	545	556	543	515	571
North Carolina	537	547	537	515	548
Indiana	522	528	520	498	545
Colorado	518	527	510	505	546
Connecticut	518	521	512	490	540
Florida	513	517	513	499	528
California	*493*	*492*	509	*454*	*495*
Alabama	*466*	*463*	*471*	*433*	*480*

*Scores are in boldface when states are performing significantly ($p < 0.05$) above the U.S. average, in regular typeface when performing not significantly differently from the U.S. mean score, and in shaded italics when performing significantly beneath the U.S. mean score (Provasnik et al. 2013).

An examination of the trend in overall performance from 2007 to 2011 in mathematics content areas for eighth graders showed no significant trend growth for the U.S. students in performance in any of the content areas from 2007 to 2011. Of the top seven countries considered similar to the United States, two did not participate in 2007, so it was impossible to check their trend. Overall, the United States showed more growth, on average, than the other countries in its similar performance group from comparisons with overall performance to content-area performance.

In examining the movement of these countries, one notes that, in general, they remain relatively stable in their groups of performing at levels significantly above, not significantly different from, and significantly below the United States. However, when students have the opportunity to learn content because mathematical content is taught before or during the year of the test, their performance may move up, and when they lack the opportunity because the content is taught in a subsequent year, their performance may move down. Curriculum and the position of mathematical content in it are major factors in determining the ranking of schools. If TIMSS were given at every grade every year, one would have a better map of this effect of the positioning of content and the resulting opportunity to learn.

Some of this variation can be seen in the inter-country variation of performance on subtests in the U.S. states' performance. However, one must also remember that other factors are involved as well. The timing of the 2011 test was such that the Common Core State Standards movement had literally no impact on the variation existing between the states, and no opportunity to lessen it, since these standards were not released until June 2, 2010.

At the eighth-grade level, there was no statistical difference between the performance of U.S. male and female students on TIMSS 2011 mathematics. Across all 56 educational systems (countries, portions of countries, and city-states) that participated in TIMSS 2011, 8 systems had males outperforming females, 35 systems showed no statistical difference, and 13 systems had females outperforming males. An examination of the differences in the nine U.S. states participating in TIMSS 2011 found that in Indiana the male students' mean performance was 8 scale points higher than the female students' mean performance—a statistically significant difference. In the other eight states, scores showed no significant differences between boys' and girls' performances at the grade 8 level.

OECD Programme for International Student Assessment Studies

The Programme for International Student Assessment (PISA), initiated in 2000, is an international assessment measuring the performance of 15-year-olds in reading, mathematics, and science literacy, using problems largely situated in context. The program is sponsored by the Organisation for Economic Cooperation and Development (OECD), largely working through committees whose members represent its 34 member nations and a group of contractors who carry out the logistical, scoring, and statistical analyses of the data (OECD 2016). The program also has special studies (problem solving in 2003 and 2012, financial literacy in 2012) that gather data on special literacy areas. Starting with 2015, problem solving will become a regular area on which data are collected. The assessment is modeled on a three-year rotating cycle. Reading, mathematics, and science are the major domains measured, and these domains rotate every three years, with the major domain for a given assessment year receiving approximately two-thirds of the assessment time for that year, with the remaining time spread among the other areas being tested. The assessment of the two non-major domains in an assessment year is limited to the number of items necessary to maintain the trend line for the two domains in their non-major-domain years.

The OECD is an international organization consisting of 34 member democracies with market economies, including Canada and the United States. These countries share information and compare and contrast their policies, experiences, and goals, with special emphases on economic growth, prosperity, and societal development. Central to these goals are inputs and outputs related to the education and resulting capabilities developed in the youth of each nation.

As such, each of the OECD countries is interested in the future composition and qualifications of its workforce, as well as sustaining its resources while improving the living conditions of its citizens. These assessments, given to 15-year-old students near the end of their compulsory education in many of the OECD countries, are less an achievement test of facts learned than an assessment of how prepared these students are to use their school- and life-acquired knowledge in real-world contextual situations (OECD 2014). Mathematics was the major domain first in the 2003 assessment, repeated as the major domain in 2012, and will repeat again as the major domain in 2021. In 2012, the OECD PISA program added a problem-solving domain to its assessment program. At this point, it is not clear in exactly what way this domain will be factored into the proportion of time allocated to domains in the assessment program's allocation of shares for particular emphases.

PISA 2012

In 2012, PISA assessed students in the 34 OECD countries as well as in a number of other education systems. The major domain of the assessment was mathematical literacy, with trend studies for reading and science literacy. Data were also collected on student performance in problem solving and financial literacy in special studies (OECD 2013a). In total, the study presented data from the 34 member states of OECD and 31 non-OECD educational entities referred to as *economies* by the OECD in reporting the results. These economies consist of countries moving toward membership in OECD, developing countries, and city-states of large nonparticipating countries. In addition, data were collected within participating countries for subregions of the countries, including three state public school systems in the United States: Connecticut, Florida, and Massachusetts. These subregions are reported, but not included in the analyses of the OECD members or the 34 OECD members plus the 31 economies. Results for the participating U.S. states are included as information to show the variation in performance at state levels within the United States. PISA scale scores are reported on a scale of 0 to 1000, with a mean of 500 (OECD 2013b).

Mathematical literacy

U.S. 15-year-olds' average score of 481 on the 2012 PISA assessment was significantly lower than the OECD average country mean score of 494. The average mathematics literacy score in the United States was lower than the average in 21 of the 33 other OECD countries, higher than the average in 5 OECD countries, and not measurably different from the average in 7 OECD countries. Korea was the OECD country with the greatest mean country score, 554. Even so, Korea was 5th among the 64 participating OECD countries and economies. Shanghai−China (613), Singapore (573), Hong Kong−China (561), and Chinese Taipei (560) all ranked numerically above Korea. However, when the analysis takes intervals of statistical significance into account, Korea emerges statistically as the highest-ranking OECD country. Nonetheless, when one examines Korea statistically alongside the 4 economies with higher means, Shanghai−China is a statistical number 1, followed by Singapore as an undisputed statistical number 2. However, looking at the intervals determined by the standard errors surrounding each of the next 3 participants, Korea could be ranked as high as 3 or as low as 5 without violating the meaning of "no significant statistical difference" existing among the 3 participants (OECD 2013b).

In 25 of the 34 OECD countries, males outperformed females in mathematics literacy. In the United States, however, the average score of males (484) was not measurably different from that of females (479).

OECD performed an analysis of the annualized change in country performance throughout OECD countries and economies participating in PISA mathematics. This analysis process resulted in a +0.3 point annualized change in U.S. performance over its participation in PISA mathematics assessments (four assessments). Turkey had the highest annualized gain, with a +3.2 point gain per year over its participation in four PISA mathematics assessments. As a comparison, the mean OECD country annualized gain was −0.3 points per year. The lowest annualized gain score for an OECD country participating in four assessments was −3.3 score scale points per year (OECD 2013b). As a result, the graph of annualized change for the United States is one of little to no change and an essentially flat trend line from 2003 to 2012.

The PISA mathematics assessment reports scores in both scale scores and achievement levels. The achievement level structure for PISA mathematics can be seen in the OECD reports for the mathematics assessments. For our purposes, we give the abbreviated achievement level intervals as follows, along with the percentage of students scoring at or above these intervals relative to the OECD averages:

- **Level 6** (scores 669 and up, describing 3.3% of OECD students): At Level 6, students can conceptualize, generalize and utilize information based on their investigations and modeling of complex problem situations and use this knowledge in relatively non-standard contextual situations. They can link different information and representations and flexibly move among them. Students at this level are capable of advanced mathematical thinking and reasoning. They can use this insight along with mastery of symbolic and formal mathematics to develop new approaches and strategies for attacking novel situations. These students are capable of reflecting on their actions and can communicate their actions and reflections precisely, noting findings, interpretations, arguments, and their appropriateness for the situation at hand.

- **Level 5** (scores 607 to 668, describing 12.6% of OECD students): At Level 5, students can develop and work with models for complex situations, identifying constraints and specifying assumptions. They can select, compare, and evaluate appropriate problem-solving strategies for dealing with complex problems related to these models. Students at this level can work strategically, using broad, well-developed thinking and reasoning skills, appropriate linked representations, symbolic and formal characterizations, and insight pertaining to these situations. They begin to reflect on their work and can formulate and communicate their interpretations and reasoning.

- **Level 4** (scores 545 to 606, describing 30.8% of OECD students): At Level 4, students can work effectively with explicit models for complex concrete situations that may involve constraints or call for making assumptions. They can select and integrate different representations, including symbolic, linking them directly to aspects of real-world situations. Students at this level can utilize their limited range of skills and can reason with some insight, in straightforward contexts. They can construct and communicate explanations and arguments based on their interpretations, arguments, and actions.

- **Level 3** (scores 482 to 544, describing 54.5% of OECD students): At Level 3, students can execute clearly described procedures, including those that require sequential decisions. Their interpretations are sufficiently sound to be a base for building a simple model or for selecting and applying simple problem-solving strategies. Students at this level can interpret and use representations based on different information sources and reason directly from them. They typically show some ability to handle percentages, fractions, and decimal numbers, and to work with proportional relationships. Their solutions reflect that they have engaged in basic interpretation and reasoning.

- **Level 2** (scores 420 to 481, describing 77.0% of OECD students): At Level 2, students can interpret and recognize situations in contexts that require no more than direct inference. They can extract relevant information from a single source and make use of a single representational mode. Students at this level can employ basic algorithms, formulae, procedures, or conventions to solve problems involving whole numbers. They are capable of making literal interpretations of the results.

- **Level 1** (scores 358 to 419, describing 92% of OECD students): At Level 1, students can answer questions involving familiar contexts where all relevant information is present and the questions are clearly defined. They are able to identify information and to carry out routine procedures according to direct instructions in explicit situations. They can perform actions that are almost always obvious and follow immediately from the given stimuli. (OECD 2013b p. 61)

As always in such systems of achievement levels, there are students who do not perform well enough to reach the level of performance described by the lowest level in the system. In the case of the OECD countries, the countries with the largest percentages of students below level 1 are among the least developed countries in the organization.

The data in table 14 provide the percentages of students, from "below level 1" through each of the 6 achievement levels, contained in each level of the achievement levels for the mean OECD country, Korea, and the United States.

Table 14

Percentage of students per country per achievement level

Achievement level	Below level 1	Level 1	Level 2	Level 3	Level 4	Level 5	Level 6
OECD mean	8.0	17.9	26.3	23.3	15.8	6.6	2.2
Korea	2.7	6.4	14.7	21.4	23.9	18.8	12.1
United States	6.0	15.0	22.5	23.7	18.1	9.3	3.3

Data from OECD, *PISA 2012 Results: What Students Know and Can Do: Student Performance in Mathematics, Reading, and Science*, Volume I. Paris: OECD Publishing, 2013b.

An examination of the data points to the fact that 44.5% of U.S. students are "below level 1" or at achievement levels 1 and 2—a situation that explains a great deal about the performance of U.S. students on the PISA 2012 mathematics assessment. It may also reflect the overemphasis of procedural correctness and the lack of attention to developing reasoning skills that marked the U.S. mathematics curriculum in the developmental years of the students in the 2012 U.S. sample.

Mathematical processes and content in PISA assessments

Such an explanation as the foregoing for the lack of achievement of U.S. students is premised on the lack of curricular content and related instruction focused on reasoning and problem-solving processes. The PISA 2012 mathematics assessment was scored to evaluate student performance on process scales. The processes defined in the *PISA 2012 Mathematics Assessment Framework* were—

> • *formulating situations mathematically;*
>
> • *employing mathematical concepts, facts, procedures, and reasoning*; and
>
> • *interpreting, applying, and evaluating mathematical outcomes* (OECD 2013a).

These processes were represented in PISA 2012 in a 1:2:1 ratio respectively.

The processes as defined in the framework are evaluated through the structuring of items and the representations used in the problem sets and then through the scoring rubrics used to evaluate students' work on the mathematical problems presented. The problems that students encountered on the assessment were divided according to the relative stress placed on content areas viewed as relevant to the development of mathematical literacy in 15-year-old students in the OECD countries. In the 2012 assessment, the content categories were *quantity, change and relationships, space and shape,* and *uncertainty and data.* Problems containing concepts, generalizations, and procedural skills associated with these areas, as outlined in the framework, were distributed in a 1:1:1:1 ratio.

Proficiency on the PISA process subscales

Each of the three process categories relate to three parts of the PISA modeling cycle. As a result, each item in the assessment was linked with one of the process categories listed above. Often part of solving a problem in context is restructuring the problem so that it can be approached mathematically. This calls for the process *formulating situations mathematically.* For a problem focused on this process, students may need to do the following:

> … recognize or introduce simplifying assumptions that would help make the given mathematics item amenable to analysis. They have to identify which aspects of the problem are relevant to the solution and which might safely be ignored. They must recognize words, images, relationships or other features of the problem that can be given a mathematical form; and they need to express the relevant information in an appropriate way, for example in the form of a numeric calculation or as an algebraic expression. This process is sometimes referred to as translating the problem as expressed, usually in real-world terms, into a mathematical problem. (OECD 2013a, p. 79)

In other settings in which the where the problem is already formulated, the challenge may be to use subprocesses associated with *employing mathematical concepts, facts, procedures, and reasoning,* and once students have recognized the basic relationships underlying the problem,

> … to carry out a calculation, substitute values into a formula, solve an equation, or apply their knowledge of the conventions of graphing to extract data or present information mathematically. (OECD 2013a, p. 83)

In yet other situations, problems may come formulated, and students are called on to judge which "solution" proposed may be the correct one or the "best" for the context used in presenting the problem. In these cases, students may be required to make

> … links between the outcomes and the situation from which they arose. For example, in a problem requiring a careful interpretation of some graphical data, students would have to make connections among the objects or relationships depicted in the graph, and the answer to the question might involve interpreting those objects or relationships. In a problem about travel on public transport or riding a bicycle, once

the basic relationships underlying the problem have been understood and expressed in a suitable mathematical form, the required mathematical processing has been carried out, and results generated, the student may need to evaluate the results in relation to the original problem, or may need to show how the mathematical information obtained relates to the contextual elements of the problem. (OECD 2013a, p. 83)

Such problems call on students' skills in *interpreting, applying, and evaluating mathematical outcomes.*

PISA then sorted the problems according to whether the major focus of the process that students would use in addressing the requirements of the problem was formulating, employing, or interpreting. For problems in each category of process, a proficiency subscale with defined levels was then constructed with levels by subscale score defined in terms of actions associated with the particular process and the relationship of subprocesses to achieving a successful application of the process targeted in a given problem and its context. A full description of the subscales construction and interpretation is contained in *PISA 2012 Results: What Students Know and Can Do: Student Performance in Mathematics, Reading, and Science*, Volume I (OECD 2013b, pp. 79–94). The PISA 2012 study report can be downloaded from the OECD PISA website: http://www.oecd.org /pisa/keyfindings/pisa-2012-results-volume-I.pdf.

The results of establishing the subscales for each process and fitting the distributions of scale scores for each country's students to the process score scale determine and describe the process levels in conjunction with the establishment of the related cut points ties to main features of the process.

This then allowed placing each country's performance relative to the process on the subscale, as was done with the mathematics achievement earlier. In the process usage analysis, the three scales were formulating, employing, and interpreting. As in mathematical achievement, the subscales ran from 0 to 1000, with a mean of 500 and a standard deviation of 100 points. The OECD average score comes from finding the mean value of the averages of the 34 OECD countries relative to a given PISA process. Because it is the mean of country averages, it may deviate somewhat from 500, which represents the level at which it is statistically estimated that all students in OECD countries would perform.

Table 15 contains the values associated with the OECD average for each subprocess and the processes as a whole, the U.S. mean for each process and rank among the 34 OECD countries, and the top performing OECD countries, their scores, and ranks.

In analyzing the data in table 15, one may wish to keep in mind that the OECD average proficiency in mathematics was 494 on the score scale. Comparing the mean scores for the formulating, employing, and interpreting subprocesses with the mean score for mathematics does not seem, on average, to signal a big difference. However, if one examines the differences of the scale score performances in mathematics with the performances in each of the three process subscales for the top six scoring OECD countries in mathematics, a different story emerges. Table 16 provides the performance scores for mathematics and the difference between that score and the respective scores in the three processes' subscales for each of the six countries and also for the U.S. scores.

Table 15

Descriptive values for mean scale score, U.S. scale score, and top OECD countries' scale scores for the mathematical processes

Formulating				
OECD mean scale score	**U.S. mean scale score**	**U.S. OECD rank**	**Top OECD countries by process scores**	**Top OECD countries' ranks**
492	475	27 out of 34	Korea (562)	No significant difference in formulating
			Japan (554)	
Employing				
OECD mean scale score	**U.S. mean scale score**	**U.S. OECD rank**	**Top OECD countries by process scores**	**Top OECD countries' ranks**
493	480	28 out of 34	Korea (553)	1
			Japan (530)	2 to 4
			Switzerland (529)	2 to 4
			Estonia (524)	3 to 5
Interpreting				
OECD mean scale score	**U.S. mean scale score**	**U.S. OECD rank**	**Top OECD countries by process scores**	**Top OECD countries' ranks**
497	489	26 out of 34	Korea (540)	1 to 2
			Japan (531)	2 to 5
			Switzerland (529)	2 to 5
			Finland (528)	2 to 5

Data from OECD, *PISA 2012 Results: What Students Know and Can Do: Student Performance in Mathematics, Reading, and Science*, Volume I. Paris: OECD Publishing, 2013b.

Examining the differences between the respective subprocess scores from the six high-scoring countries and their mathematics scores indicates that most did slightly better on *formulating* problems than they did on their mathematics examination overall. On *employing*, their subscores were closer to their mathematics scores, but slightly lower, indicating that the specific focus items on *employing* knowledge might have been slightly more difficult than the average mathematics item. The results for the comparison of the mathematics score indicate that they might have been slightly more difficult than the other subprocess items.

For the United States, performance differences between the U.S. mathematics score and the subprocess scores showed a different profile. U.S. students seem to have found the *formulating* items more difficult than the mathematics items overall; the *employing* items about the same, on average, as the mathematics items; and the *interpreting* items much easier than the mathematics items overall. While relatively balanced, the profile tends to indicate that U.S. students may have developed understanding and the ability to interpret and compare results at a higher level than average OECD students, although

Table 16

Performance difference between the overall mathematics scale score and each process subscale score

Country	Mathematics score	Formulating	Employing	Interpreting
Korea	554	8	−1	−14
Japan	536	18	−6	−5
Switzerland	531	7	−2	−2
Netherlands	523	4	−4	3
Estonia	521	−3	4	−8
Finland	519	0	−3	9
U.S.	481	−6	−1	8

Data from OECD, *PISA 2012 Results: What Students Know and Can Do: Student Performance in Mathematics, Reading, and Science*, Volume I. Paris: OECD Publishing, 2013b.

at the same time, they may have failed to develop their formulation process skills to the same degree.

The overall analysis of these difference scores across the 34 participating OECD countries shows the mean difference between their highest and lowest difference scores to be about 5 scale points. Six OECD countries show their highest mean process score in *formulating*, 10 in *employing*, and 18 in *interpreting*. Korea, Japan, Switzerland, the Netherlands, Iceland, and Turkey had their highest process scale scores in *formulating*, and four of these countries are the best-performing OECD countries on the mathematics assessment.

Proficiency on the PISA content subscales

In a like manner, PISA carried out an evaluation of student performance in the four mathematical subcontent categories *quantity, change and relationships, space and shape, and uncertainty and data* (OECD 2013a). As mentioned earlier, the subcontent category items were equally distributed among the items on the 2012 PISA mathematics assessment. By means of methods like those used on the mathematics assessment and the analysis of the process items, the items in the mathematics content subcategories were analyzed and described in their content in even more detail, and a performance subscale reflecting the content of the items was developed. The PISA 2012 report mentioned earlier details the performance subscales for the content sub-categories and describes the development of their achievement scales.

We begin our analysis of the findings with the *quantity* subscale category. The average OECD subscale mean performance for the *quantity* content area was 495. This value was 1 scale point higher than the OECD mean performance for the total mathematics test. The top performing OECD countries for this subscale were Korea (537), the Netherlands (532), Switzerland (531), and Finland (527). From a statistical ranking sense, Korea ranked from 1st to 3rd place, the Netherlands and Switzerland from 1st to 4th place, and Finland from 3rd to 5th place. Hence, Korea performed statistically higher than Finland, but not significantly differently from the other three in this top grouping on *quantity*. Likewise, there were no significant differences among the performances on *quantity* items for the other three members of the group. The U.S. performance on the *quantity* subscale was 478, a score 3 points below its total mathematics scale score of 481. This subscale score singled the U.S. out for the rank of 26th among the 34 OECD countries.

The analysis of OECD students' performance on the *change and relationships* subscale found a mean OECD performance of 493 points. The high-performing OECD countries on this subscale were Korea (559), Japan (542), Estonia (530), and Switzerland (530). From a statistical ranking standpoint, Korea ranked number 1, statistically higher than the remaining OECD countries in performance on *change and relationship* subscale items. Japan ranked number 2, a position statistically below Korea and statistically higher than the remaining 28 OECD countries. Estonia ranked from 3rd to 4th position, while Switzerland's ranking interval was from 3rd to 5th position. Because these ranking intervals overlap, there is no statistical difference between the ranking positions of these two countries. The U.S. performance for *change and relationships* was 488, 7 points higher than its overall mathematics assessment result of 481. Its statistical ranking interval out of the 34 OECD countries was from 18th to 24th position.

When the subscale performances were analyzed for *space and shape*, the OECD subscale mean performance was found to be 490 points, a level 4 points below the OECD mean for this assessment test of 494. The top four OECD performers were Korea (573), Japan (558), Switzerland (544), and Poland (524). Statistically, these four OECD countries were ranked 1, 2, 3, and 4, respectively, with statistically significant differences between their respective performances on the subscale assessment. The U.S. performance on the *shape and space* subscale was 463, a performance that was 18 score scale points below its performance on the overall OECD mathematics test.

The fourth and final mathematics content subscale was for *uncertainty and data*. This subscale had a mean OECD scale score of 493 points, 1 point lower than the OECD overall mathematics achievement assessment. The top performing OECD countries on this content subscale assessment were Korea (538), the Netherlands (532), Japan (528), and Switzerland (522). Statistically, the analysis of ranking differences for these 5 countries resulted in Korea ranking 1st to 2nd, the Netherlands ranking 1st to 3rd, Japan ranking 2nd to 4th, and Switzerland ranking 3rd to 6th. Thus, there was no significant difference in the subcontent area of *uncertainty and data* for Korea, the Netherlands, and Japan, nor was there any significant difference for the Netherlands, Japan, and Switzerland. However, there was a significant difference in performance for this subcontent subscale for Korea and Switzerland, with Korea's performance ranking significantly above that of Switzerland for *uncertainty and data*. The U.S. subscale score for this subcontent area was 488. That resulted in an OECD statistical ranking for the United States somewhere in the interval from 19th to 26th place out of the 34 participating countries.

The overall mathematics score for the OECD assessment correlates highly with the scores from the mathematical content subscale score results. But the differences in subscale scores among the OECD countries vary greatly, and the variation is even greater than the process subscale variation discussed previously. *Space and shape* tends to be seen as the easiest content subscale among the high-performing countries, and *change and relationships* as one of the most difficult for low-performing countries.

An examination of OECD countries' differences between the subscale for their easiest content area by subscale and their most difficult content area by subscale area shows that Japan has the largest difference between its strongest (*space and shape*) and weakest (*quantity*) subscale content areas of 39 points; Turkey has the smallest difference between its strongest and weakest subcontent areas of about 7 points. This is almost identical to Turkey's subscore area differences in the subprocesses analysis. Between the extremes of these two countries is a great spread, with an average difference between a country's strongest and weakest performance being about 17 points. For the United States, the gap between strongest and weakest performances on the mathematical content subscales was between *change and relationships* and *uncertainty and data*, tied for easiest at 488 (7 points above the U.S. mean math performance of 481), and *space and shape*, at 463—17 points below the nation's overall mathematical assessment average of 481. This gives the U.S. a variation of 25 points between the strongest and weakest subscale scores (OECD 2013b).

The foregoing analyses provide a picture of U.S. performance on the TIMSS and PISA assessments, with a special emphasis on the 2011 and 2012 assessments, respectively. TIMSS, and the IEA tests in general, focus on student achievement in areas where they are expected to perform well—areas in which, on average across the countries, they have had an opportunity to learn. PISA, along with other tests that the OECD is involved in administering and analyzing, focuses more on what students can do with what they have learned up to the 15-year-old stage of life. This learning may have come not only through schooling but also through life experience in the community, self-study, or a variety of other factors. Finding these differences in similar-aged samples causes one to ask, What have we learned?

U.S. Gains from the TIMSS and PISA Assessments

What has the United States gained from the TIMSS and PISA assessments? This question has been posed in many ways in various settings and hearings over the past 60 years of the IEA studies and 15 years of the PISA studies. It is a valid question, in that these studies require millions of dollars to design and mount on the international educational front. The results, however, are among the most prominent gauges of educational achievement in the participating countries. One of the authors recalls when the results of SIMS were first released, a minister of education from one of the participating countries who was present in the room said under his breath, "There will be trials."

Given the costs and the public prominence of the results and comparisons among countries, many of which are inappropriate or statistically unsound, there must be counterbalancing gains—outcomes that provide information that can assist in the strengthening of curricula on an individual country level for participating nations and broader patterns of understanding for the participating countries as a whole. Any approach to the data emanating from a large-scale assessment such as TIMSS or PISA requires a knowledge of the context in which education occurs in a given participating country and the cultural expectations for schooling outcomes held by its citizens and students. For example, are the assessment tests viewed as high-stakes tests or low-stake events that students can "blow off?" Or how well do the items on the assessments correlate with the curriculum with which participating students have had experiences up to the time of the assessments? Or to what degree have students in the participating countries had opportunities on a regular basis to participate in assessments asking them to answer multiple-choice questions, short constructed-answer questions, and extended constructed-response questions with scaffolded subquestions, graded by a multipoint rubric? Or to what degree have students had access to technology that assists them in solving problems of the type that appear on the examinations and answering the questions by keying in their responses? All of these and other factors need to be addressed at the beginning of any evaluation of the outcomes produced by these studies.

We begin our analysis of some outcomes with the FIMS test and then move to the SIMS and finally to the Third International Mathematics and Science Study of 1994–95. These studies provide a useful entry point to answer the foregoing questions and to point out other factors that weigh heavily in the analysis of IEA, and similar studies.

Outcomes of the first IEA studies

The first notice that the public—or most American mathematics educators—had about the existence of the First International Mathematics Study appeared in the *New York Times* on Sunday, March 12, 1967, when the *Week in Review* section released the findings of the study and the United States' position in the rank-order finish of the "horse race" of nations (Hechinger 1967). This article and the emphasis on which nation was in first place, which was in second, and that the United States was last out of 13 nations provoked a great deal of consternation about the state of mathematics education in the United States. It fueled the debates about the "New Math" and did little to provide evidence showing what the study was about.

FIMS: The First International Mathematics Study

The most important finding of the First International Mathematics Study (FIMS) was the role that opportunity to learn, or OTL, plays in determining children's achievement levels. The report of the FIMS examination of this variable showed strong correlations between the amount of exposure that students have to mathematical content and their capability to perform on items related to that content (Grouws and Cebulla 2000; Husén 1967).

FIMS also exposed the great difference in goals and foci of the curricula of the 13 nations that participated in this early international study. Central to the study was what the study directors and IEA learned about mounting an international comparative study of mathematics curricula and the associated student achievement. Although all studies can do better, FIMS showed that the sampling frames, especially for the students in the last year of secondary school, cut out a large number of U.S. students in that very class because of the way in which the criteria were structured, mainly as a result of a lack of understanding of the U.S. curriculum at that level.

SIMS: The Second International Mathematics Study

SIMS took place in the Northern Hemisphere at the end of the 1981–82 school year. Along with a number of other countries, the United States tested students at the beginning of that school year as well, to obtain a marker for growth across the school year, instead of relying on the end-of-year mark to describe the totality of the students' school year.

Students from two populations were sampled. The first, population A, consisted of students in grade 8 in the United States and most other countries but students in grade 7 in Japan and Hong Kong. The U.S. students were slightly above the international average in computational arithmetic and comparable to the international average in algebra. They were below the international average in the other areas tested and well below in non-computational arithmetic—problem solving. The U.S. eighth-grade students were slightly below their mean for the FIMS tests. The declines were somewhat greater for the more demanding comprehension and application items than for computational items (McKnight et al. 1987).

The second group of students tested, referred to as population B, consisted of twelfth graders taking college preparatory mathematics courses. These courses ranged from trigonometry, precalculus, and analytical geometry to calculus. The last course was divisible into an introductory concepts course and Advanced Placement courses. The average achievement of the U.S. students in population B was substantially below the average in the other countries when a weighted average across the content areas was used. The average achievement of U.S. students taking calculus was at or slightly below the average of the advanced secondary school mathematics students in the other countries. Thus, the best (~20%) of U.S. students in their last year of secondary college preparatory mathematics were at or below the average for all the last-year students in other countries. The achievement of all U.S. precalculus students was also substantially below the average of all students in the other countries.

Challenges in mathematics for population A

An examination of the coverage of topics in the framework for the population B achievement tests showed that the opportunity to learn the material in U.S. classrooms was typically below the average OTL in other countries. However, the OTL varied among classrooms and among schools in the U.S. sample, even for students taking the same level of classwork.

Yet, the most startling finding—especially for those not involved in the teaching of mathematics—was the differentiation of opportunity to learn by class types at the population A level in U.S. classrooms. At the eighth-grade level, there were four distinct tracks of curriculum, which could be labeled as remedial, typical, enriched, and algebra 1. With respect to the textual materials they used, these four tracks might be referred to as refresher arithmetic, grade 8 mathematics, pre-algebra, and algebra 1, respectively. However, at the end of grade 8 in the United States, the students in the lower three tracks may, at some schools, all be placed into algebra 1 as they enter high school, since some high schools do not accept the algebra courses taught at the junior high or middle school grade 8 level. But by the end of the grade 8 year, the students in the remedial, typical, and enriched tracks had learned 24%, 38%, and 58%, respectively, of the content of the algebra on the SIMS test. At the beginning of the same school year, the algebra 1 students already had the knowledge required for 56% of the items on that test. This shows the great variability in the students who will, most likely, be mixed together in algebra 1 of some form in their first year of high school. In fact, each of the three lower groups had attained in grade 8 what the next higher group already knew at the beginning of their eighth-grade year!

One might argue that such delineation of students to the varied tracks perhaps increased efficiency and instructional effectiveness. However, a variance decomposition of the factors that accounted for student achievement growth—school, class, or student for the students in the population A sample for the United States—found that the major factor was the curricular differentiation between schools, and especially between classes within schools, that set the boundaries on what students had the opportunity to learn. The amount of variation in academic growth for students was very low—less than that in any other country (McKnight et al. 1987; Burstein 1993; Kifer 1993). The data show that many students in the United States are being deprived of opportunities to participate in the best classes in mathematics that their schools have to offer as a result of a sorting mechanism that governs how they move from one class level to another.

The U.S. portion of the study, relative to the U.S. data contrasted with that of other countries, demonstrated the need for—

- a restructuring of the mathematics curriculum;

- a reexamination and revitalization of the content of the mathematics curriculum, even if restructured;

- the establishment of clear standards for achievement at each grade level to create an institutionalized climate of expectation to which students will respond; and

- a reexamination of the practice of early sorting of students into curricular tracks that lead to vastly different opportunities to learn high school mathematics (McKnight et al. 1987, p. 130).

A comparison of these 1987 recommendations with the outcomes of the move to standards spearheaded by various state and professional organizations and the lessening of the amount of tracking within schools suggests that the findings of international studies and comparisons of differing curricular models do have an influence on the curricular decisions made by countries.

Challenges in mathematics for population B

Many of the recommendations for population A indicate direct lines of action for the beginning courses in secondary school mathematics and hence are recommendations for secondary schools as well. The call for strong new curricula and standards also applies at this level. The factors behind low enrollment in the most advanced mathematics classes available in the United States need to be studied and ameliorated.

An examination of the population B findings shows that the big factor still remaining is that the mathematics curriculum is the distributor of opportunity to U.S. students. A consensus must be reached in two areas:

- A recognition of what students need to learn and what they need to be able to do with it when they have learned it

- A commitment to giving more students access to the path to more challenging and rewarding study of mathematics

Until agreement and resolution are achieved in these areas, little change is likely in the picture of national achievement findings.

Although SIMS substantiated many of the hypotheses that were stated at the end of FIMS, it lacked the data to strengthen them to hard findings. However, SIMS created data-driven hypotheses, whereas FIMS only suggested the same relationships. These findings from SIMS provided a direct link to framing instrumentation for such studies in TIMSS and curricula and instructional methods that promised to provide more instructional and assessment programs a decade later (Travers and Westbury 1989; Robitaille and Garden 1989; Burstein 1993; McKnight et al. 1987). Further, while not directly driving the development of the NCTM Standards (1989), the findings of this SIMS study were known to the leaders of the NCTM Standards movement and clearly had substantial influence on decisions made and directions taken (NCTM 1989; Dossey and Lindquist 2002).

Ongoing comparative studies of mathematics education: TIMSS and beyond

TIMSS, initially the Third International Mathematics and Science Study, has now morphed into the Trends in International Mathematics and Science Study in the hands of IEA and has become a series of studies building on a trend line like NAEP. It has designed a regular cycle of assessments appearing every four years, with increasing involvement of participants in the mathematics test and its partner PIRLS, Progress in International Reading Literacy Study to document progress in mathematics and reading achievement. The TIMSS/PIRLS International Study Center at Boston College envisions the participating countries using TIMSS and PIRLS to

> ...explore educational issues, including: monitoring system-level achievement trends in a global context, establishing achievement goals and standards for educational improvement, stimulating curriculum reform, improving teaching and learning through research and analysis of the data, conducting related studies (e.g. monitoring equity or assessing students in additional grades), and training researchers and teachers in assessment and evaluation (TIMSS & PIRLS International Study Center 2015).

TIMSS: The Third International Mathematics Study

TIMSS took place in 1994–95, with data collected at five grade levels—grades 3, 4, 7, and 8, and the final year of secondary school—in more than 40 countries. The international report provided data on education policy in the countries and how decisions were made about education, how textbooks and other materials were selected for classroom use, and how examinations were used (Mullis et al. 1997, 1998; Beaton et al. 1996). For U.S. performances in TIMSS, the National Center for Education Statistics (NCES) issued a set of three reports in 1995, one for each of the three levels 4, 8, and the final year of secondary school. These reports filtered the U.S. data from the international report and provided an interpretive view of U.S. performance against a matrix of other participating countries. The reports were all entitled "Pursuing Excellence: A Study of U.S. XXX-Grade Mathematics and Science Achievement in International Context" (1995), with the "XXX" replaced by "Fourth," "Eighth," and "Twelfth," in succession. Lois Peak and NCES colleagues authored the volumes for the grade 4 and grade 8 publications (1995a and 1995b) and Sayuri Takahira, Patrick Gonzales, Mary Frase, and Laura Hersh Salganik authored the grade 12 volume (1995).

The unique new feature in the TIMSS study was a video study on the teaching of mathematics in six different countries: Australia, the Czech Republic, Hong Kong SAR, the Netherlands, Switzerland, and the United States. The video study was expanded to seven countries when the videos taken of Japanese mathematics classrooms as part of a TIMSS study in 1995 were recoded and included as part of the data set.

This study was actually carried out in 1999 as part of a repeat of TIMSS 1995 to study the growth of students from the same general population as in the 1995 TIMSS assessment. These videos were made to delve deeper into the similarities and differences of classroom instructional methods in the countries. One of the first differences observed was in the certification of the teachers in the various countries. When the teachers in the videos were asked whether they were certified to teach eighth-grade mathematics, 79% of the U.S. eighth-grade teachers reported that they were, while the percentages reported in other countries were as follows: Australia, 66%; Czech Republic, 85%; Hong Kong, 100%; the Netherlands, 97%; and Switzerland, 48%. The only significant differences ($p < 0.05$) were that the Netherlands had a significantly higher rate of certified teachers for grade 8 than Australia, and the U.S. had a significantly higher rate than Switzerland. Other U.S. teachers reported being certified for teaching science or some other grade level of schooling. U.S. teachers had a mean of 14 years of experience, ranking them numerically in the middle of the seven countries. However, the U.S. teachers' mean experience was significantly less than that of teachers in the Czech Republic, with 21 years of experience.

An examination of the goals that teachers had for their lessons found no significant differences between the goals of U.S. teachers and those of teachers from other countries for percentage of lessons with a focus goal on content (81%), process (96%), or perspective (11%). Content goals focus on specific concepts or topics, process goals focus on describing how to solve or apply topics discussed, and perspective topics were devoted more to developing an interest in mathematics as a field and promoting good attitudes about mathematics.

The report provided a great deal of information and has to be read in depth to ascertain what it means for a given country. Overall, it indicates that different patterns of instruction are effective for different countries, and more than one of these can be associated with high achievement. Hong Kong and Japan, two high-scoring countries, spent 76% of their time on working with new content and 24% of the time on reviewing past content. However, teachers in Japan spent more time introducing the lesson, and teachers in Hong Kong devoted more time to practicing the new content. During the lesson, teachers in Japan spent more time on trying to get students to make connections, while teachers in Hong Kong devoted more time to practicing the new material. These show the differences between two of the highest-scoring participants. The authors of the project report even indicate if one thinks a comparative study will identify the best way, one will find that the comparisons do not yield a clear answer, even among high-achieving countries.

However, similarities do emerge. All the countries involved spent at least 80% of their time working on solving problems and some lesson time on presenting new content. The ways in which they allocated the time within lessons to these activities varied. The authors of the video study project posited that a deeper understanding of how to align instruction with specific learning goals might yield more answers (Hiebert et al. 2003).

TIMSS 2004–2008: Trending into a pattern

Across the years since the period 1995–99, the TIMSS format has seen fewer adjustments and a sharper focus on examining students' processes—especially students' capabilities to communicate through reading and writing in doing mathematics—and trends in student achievement on the content domains. Although politicians like to use the results to secure press time and promote policies, the researchers involved with TIMSS have been strengthening the assessment to detect differences in student content achievement and process use. Those in the assessment community have been urging politicians and developmental foundations to avoid forcing a curricular plan that works in one country onto another that has a different culture, educational goals, and societal wishes for its children. History has shown that the adoption of a curriculum without consideration of the culture and ethnic identities tied to it may diminish or jeopardize its chances of producing the outcomes desired (Dossey and Wu 2013).

Questions arising from the field about TIMSS findings are now directed mainly at technical issues in sampling and adjusting the sample for changes in the demographic compositions of participating countries' social classes over time, issues related to the application of technology to the collection of assessment results, and other issues. For the moment, it appears that the content has stabilized, and mathematics educators are looking deeper to determine the meaning of the trend lines that are emerging.

Chapter 6: Implementing and Assessing the Common Core State Standards

Earlier chapters have described the Common Core State Standards for Mathematics (CCSSM; NGA Center and CCSSO 2010a) and the way in which they provide a common mathematics curriculum for all states in the United States that have chosen to adopt them. With an increasingly mobile population across the states and regions, having common grade-level expectations is reassuring and helpful for teachers, as well parents and students. In addition, CCSSM advocates students' development of a deeper understanding of central concepts and mathematical processes, as well as an expanded fluency with mathematical procedures.

Although most states initially adopted or adapted CCSSM as their states' curriculum standards in 2010, a few states are beginning to back away from these decisions under pressure from constituencies. One significant point of contention appears to be the accompanying change to standardized electronic assessments that have not been fully tested and are not customizable for special populations. Another issue for many states is a lack of funds and a lag in the availability of curriculum and resources for teacher support. In addition, as some districts, and possibly some states, have witnessed the attack on the Common Core State Standards, they have decided to join the debate and postpone making a change in the status quo. Whatever the reason, the result has been that three states have reversed their decision to adopt the standards, and a few other states are considering similar changes.

As previously noted, 22 states and the District of Columbia have adopted CCSSM verbatim as their schools' curriculum standards in mathematics, while another 21 states have adopted these standards with some modifications. In many of these states, curriculum guides and suggested textbooks are just becoming available to teachers, whereas standardized assessments of student learning based on these standards have already begun. For example, full implementation of CCSSM was expected by the 2014–15 academic year in 26 states, and standardized assessments were administered within these states. Although most testing results have not yet been released, a few states are beginning to publish results. These assessment results are frequently used as a tool for assessing teaching as well as student achievement of the standards, so teachers of mathematics across the United States are coming under pressure to adhere to CCSSM, despite the lack of resources and support. As a result, some states and districts have been developing their own curricula to maintain the pace of implementation of the standards, despite the fact that this way of proceeding is likely to counteract CCSSM's overall goal of having a common curriculum across districts and states. Chapter 3 provides details of commercial textbooks and resource materials being used in the various grade bands, but no data are currently readily available regarding the use of noncommercial texts or digital options. In addition, Chapter 4 provides a summary of the two commercially available assessments, SBAC and PARCC.

Although the Common Core State Standards were released in 2010, many regions of the United States are still adapting their curricula to meet these standards. Information Market Research has recently released data and analyses on current implementation of the Common Core State Standards by states (Resnick and Sanislo 2015). These indicate that of those states that have adopted CCSSM in full or with modifications as their state curricular standards in mathematics, 49% "have fully implemented [the standards], at least at some grade levels," and another "34.5% have started to implement" the standards. According to plans reported by states for phasing in CCSSM (Academic Benchmarks 2015), it may be another three to five years before we can appreciate the overall influence of these standards—or perhaps the lack of impact—on mathematics education in the United States.

Challenges to implementing CCSSM, as with any new curriculum, lie in interpreting the standards in a way that remains true to the authors' vision, identifying appropriate materials for use in the classroom, including materials for students with special needs, and providing resources to support teachers in the transition. Interpretation and availability of classroom materials—more specifically, print materials, such as textbooks—have been addressed in other parts of this report. Thus, the sections that follow highlight implementation challenges as they relate to the availability of resources and professional development for teachers and the assessment of student learning. Overall assessment of CCSSM is also discussed.

Resources for Interpretation and Implementation of the Common Core

Teachers and administrators in many states are phasing in the Common Core State Standards in Mathematics slightly ahead of the appearance of curricular materials developed specifically to address these same standards. Some publishing companies have created supplemental materials to bridge the gap in content. Even so, there is some indication that districts are beginning to be concerned about the lack of appropriate materials and in response are beginning to develop their own. In addition, the rapid growth of digital curricular materials and supporting tools are markers of needed professional development for teachers. Thus, the need for resources to guide the interpretation and development of materials for implementing CCSSM is evident. In its 2011 report *A Priority Research Agenda for Understanding the Influence of the Common Core State Standards for Mathematics*, Horizon Research identified research on the development and revision of curricular materials as one of six top priorities. In particular, the authors called for a research focus on how curriculum developers interpret the standards and what expertise and resources they use to inform the revision or development of materials (Heck, Weiss, and Pasley 2011). In what follows, we briefly examine some of these same issues about resources that are currently available for curriculum developers and teachers from varied sources in response to CCSSM. We also provide links to more expansive discussions of the issues surrounding the standards and openly available resources.

A reasonable place to begin a search of resources is the Common Core State Standards Initiative website, http://www.corestandards.org. This site provides links for reading and downloading the actual standards as well as supporting documents developed by the National Governors Association Center for Best Practices and Council of Chief State School Officers writing team. Moreover, the site provides direct links to individual states' standards sites at http://www.corestandards.org/standards-in-your-state/. These sites typically provide details on the states' implementation timelines, implementation efforts, and additional state resources.

The main CCSS Initiative website contains a few other links that are helpful for interpreting CCSSM, including the following:

- A document published by the Council of Chief State School Officers providing a look into the development of the standards, including the overall rationale and summary of the research supporting them (Conley 2014)

- An appendix illustrating ways in which the mathematics standards might be implemented in either a traditional secondary curriculum or a secondary school pursuing an integrated mathematics approach in its curriculum (http://www.corestandards.org/assets/CCSSI_Mathematics_Appendix_A.pdf);

- Publishers' guides providing criteria for the revision or development of mathematics curriculum aligned with CCSSM at the K–8 grade levels and the secondary school level. These guides emphasize three major themes of the standards: focus, coherence, and rigor (see, for example, http://www.corestandards.org/wp-content/uploads/Math_Publishers_Criteria_K-8_Spring_2013_FINAL1.pdf).

- A statement discussing equal access to a rigorous curriculum for students with disabilities so that they may be better prepared for success after high school (http://www.corestandards.org/wp-content/uploads/Application-to-Students-with-Disabilities-again-for-merge1.pdf)

Aside from the standards documents themselves, the CCSS Initiative website falls short of providing examples of ways in which the standards could be interpreted and implemented in a classroom.

Common Core Resources Developed by the National Council of Teachers of Mathematics	Initially, NCTM focused much of its efforts on demonstrating links between *Principles and Standards for School Mathematics* (NCTM 2000) and the Common Core State Standards for Mathematics. To some, this appeared to be a defensive move or a move demonstrating a lack of support for the Common Core State Standards. However, the NCTM leadership has been supportive of these new standards from the beginning, and more recently, much of NCTM's efforts have been directed toward support for teachers in implementing CCSSM. For example, NCTM leadership and membership have been working to help correct misconceptions surrounding CCSSM, such as the notion that implementation struggles (e.g., as a result of ineffective instructional materials) may be taken as a sign of poorly developed standards. In its policy statement in support of CCSSM, NCTM notes, "When properly implemented, the Common State Standards will support all students' access to, and success in, high-quality mathematics programs. Such programs lead to knowledge of mathematics content and reasoning skills that enable students to apply mathematics effectively in a myriad of careers and in everyday life" (NCTM 2013). NCTM's resources for supporting teachers and districts include professional development workshops, a wide range of publications, a video series, and multiple presentations at national and regional meetings. Many of these resources are detailed below:

- **NCTM Interactive Mathematics Institutes.** NCTM has developed and delivered various institutes across the United States over the past few years. According to NCTM, these Interactive Institutes "offer two and a half days of face-to-face, in-depth professional development provided by experts in mathematics education." Recent institutes have focused on aligning instruction with CCSSM and effective teaching strategies for implementing career- and college-readiness standards. More details can be found at http://www.nctm.org/Conferences-and-Professional-Development/Institutes/.

- *Making It Happen: A Guide to Interpreting and Implementing Common Core State Standards for Mathematics* (2011). This e-publication identifies NCTM's array of books, instructional guides, and CCSSM-related materials that can help teachers, administrators, and schools make the transition from their existing mathematics curriculum and instructional efforts to those centered on the Common Core State Standards for Mathematics. This work connects CCSSM with NCTM's major standards documents, such as *Principles and Standards for School Mathematics* (2000) and *Curriculum Focal Points for Prekindergarten through Grade 8 Mathematics* (2006). *Making It Happen* also points teachers to volumes in NCTM's *Principles and Standards for School Mathematics* Navigations Series, thirty-five books designed to present activities to support teaching aligned with NCTM's Content Standards and Process Standards in particular grade bands and grade levels.

- *Principles to Actions: Ensuring Mathematical Success for All* (2014). This publication describes actions that build on and support NCTM's Principles for promoting high-quality mathematics education for all students and the CCSSM-supported mathematics education for career- and college-readiness. *Principles to Actions* focuses on the Teaching and Learning Principles and describes "eight essential, research-based Mathematics Teaching Practices," as well as potential obstacles to effective teaching. The intended audience includes teachers, administrators, parents, and policymakers. At the present time, NCTM's Interactive Institutes are designed to elaborate on and reinforce the ideas in *Principles to Actions*.

- **Essential Understanding Series (2010–2014).** NCTM developed this series of sixteen professional development books to deepen teachers' mathematical understanding of challenging key concepts as the first step in effective teaching of such concepts. The books in this series are intended to help teachers develop their own understanding and prepare them to confidently help students do the same. The books address a wide range of topics, including number and numeration in K–grade 2, algebraic thinking in grades 3–5, proportional reasoning in grades 6–8, and statistical reasoning in secondary school. This series is seen as a particularly useful resource for teacher professional development.

- **Putting Essential Understanding into Practice Series (2013–).** NCTM designed this series to build on and expand the Essential Understanding Series by exploring best practices for teaching the fundamental concepts that students need to grasp to develop a robust understanding of mathematics. The books examine classroom vignettes and samples of student work to help teachers build their pedagogical content knowledge in alignment with CCSSM. Twelve volumes are planned for the series; eight have been published to date. Each volume pairs with a volume on the same topic in the Essential Understanding Series.

- **Implementing the Common Core State Standards through Mathematical Problem Solving (2012–2014).** This series of professional development materials was developed by NCTM to enhance teachers' capabilities to implement the recommendations of the Common Core State Standards in the areas of problem solving. Problem solving is seen as the core mathematical process in NCTM's *Principles and Standards* and has been a consistent theme in mathematics education reforms for many decades. Thus, it is not surprising that problem solving is seen as a mechanism for engaging students with the content of the Common Core.

This NCTM series of four books provides teachers with a variety of engaging and "highly effective" problems that link with the Common Core's content domains.

- **Connecting the NCTM Process Standards and the CCSSM Practices (2013).** This NCTM-sponsored publication (Koestler et al. 2013) highlights the Standards for Mathematical Practice outlined in CCSSM. CCSSM identifies eight practices but does not integrate them well with the content standards. As a result, they are overlooked or ignored in many cases. In other cases they are integrated but only through puzzle problems and number tricks. The practices need to be integrated through examples and projects that connect content and develop strong problem-solving skills supporting modeling and the structure of the mathematics underlying the CCSSM standards. Good examples can provide ideas about how to integrate the practices seamlessly into effective instruction and successful problem solving. The authors of this NCTM publication demonstrate that the CCSSM practices are not new ideas but are closely related to the core mathematical processes defined in other standards documents. By providing guidelines for incorporating the practices into the teaching of the mathematical content of the standards, this book serves as a valuable resource for teachers who are working to implement the Common Core.

- *Teaching and Learning Mathematics with the Common Core* **video series**. A joint effort of NCTM and the Hunt Institute—an affiliate of the University of North Carolina at Chapel Hill—the videos in this series are intended to demonstrate the progression of mathematical ideas from the earliest grades to college. The videos provide discussions with experts in mathematics education and highlight major concepts and ways to teach them for understanding. Teachers in the videos also demonstrate ways to incorporate the Standards for Mathematical Practice. All the videos in this series are available without charge on the NCTM website at http://www.nctm.org/Standards-and-Positions/Common-Core-State-Standards/Teaching-and-Learning-Mathematics-with-the-Common-Core/.

Common Core Resources Available through the Institute for Mathematics & Education at the University of Arizona

The Institute for Mathematics & Education at the University of Arizona hosts two important websites that are related to CCSSM and are being developed through projects that grew out of the Institute for Mathematics and Education in the Department of Mathematics. The full set of projects can be seen at http://ime.math.arizona.edu/IME_Programs.html.

- **The Illustrative Mathematics Project.** Created at the University of Arizona under the leadership of William McCallum (University of Arizona) and Kristen Umland (University of New Mexico), with initial funding from the Bill and Melinda Gates Foundation, the Illustrative Mathematics website now operates as an independent 501(c)3 nonprofit association sharing a wide variety of vetted resources in mathematics education for teachers, curriculum developers, curriculum coordinators, and district administrators. Website links give access to the following:

  o The individual CCSSM standards, sorted by various content dimensions and grade levels, with access to tasks associated with specific standards and cognitive outcomes. Commentary and possible task solutions provide additional information on possible student reasoning and common misconceptions.

o The CCSSM Standards for Mathematical Practice, with a link to a document that provides detailed elaborations of these standards for K–grade 5 and grades 6–8. Videos provide illustrations of the standards as they can be enacted in the classroom, and sample tasks and related commentary are also available for some standards.

o Course blueprints for each level, K–grade 8, highlighting the key ideas for those grade levels. The website authors also demonstrate connections to previous grade levels and make suggestions for organizing a curriculum that addresses all standards for the given grade level. At the high school level, blueprints are provided for content-specific courses, such as a course in geometry or algebra, and blueprints are also offered for an integrated sequence of mathematics courses.

o Video lectures providing elaborations and illustrations of the progression of concepts that are essential for teaching and learning fractions in grades 3–5. Sample tasks are associated with each video.

o A professional learning link providing access to an array of options for teacher professional development and other resources. Some of the options include virtual lectures hosted by mathematics education professionals from around the United States, one- to three-day conferences for schools and districts, and print materials, such as the progressions documents (described below) and sample tasks.

The project website can be found at https://www.illustrativemathematics.org. One effort that the project might usefully undertake is the development of college course materials for preservice and in-service K–12 teachers, detailing both content and pedagogical approaches to teaching the CCSSM standards in today's classrooms. The CCSSM learning progressions (described below) might play a central role in such an effort.

- **The Progressions Documents for the Common Core Math Standards Project.** Funded by the Brookhill Foundation under the leadership of Phil Daro (Strategic Education Research Partnership, San Francisco) and Jason Zimba (Bennington College), these documents focus on the development and display of the learning hierarchies undergirding CCSSM. As the authors note, the progressions documents "can explain why standards are sequenced the way they are, point out cognitive difficulties and pedagogical solutions, and give more detail on particularly knotty areas of the mathematics." The documents are intended for use in teacher preparation and professional development, as well as textbook development and curriculum organization. As of 2013, the team of writers, many of whom were involved in the writing of CCSSM, had developed drafts of learning progressions showing the flow of concepts, principles, and important landmarks that students encounter as they progress through a CCSSM-based curriculum in the following content areas or concepts:

o Draft K–5 progressions on geometry, measurement and data, number and operations (K–grade 5, base ten; grades 3–5, fractions), and algebraic thinking

o Draft 6–8 progressions on ratios and proportional relationships, the number system, statistics and probability, and expressions and equations

o Draft high school progressions on statistics and probability, algebra, functions, and modeling

These materials, and others to come, are available at http://ime.math.arizona.edu /progressions.

Additional Resources for Interpreting and Implementing the Common Core

Other resources are becoming available to support the interpretation and implementation of CCSSM. Two useful ones are identified below:

- **Edutopia: Resources for Understanding and Implementing the Common Core State Standards.** The Edutopia website was founded by George Lucas as a clearinghouse for information on what works in education. The Common Core resources at Edutopia include links to useful videos, such as those that are openly available at the Teaching Channel or the Hunt Institute YouTube channel. This website also describes current projects and links to these projects and other education-focused sites that would be useful to teachers for implementing CCSSM, including the Partnership for the 21st Century Common Core Toolkit and Achieve's Achieving the Common Core. Edutopia can be found at http://www.edutopia.org /common-core-state-standards-resources#graph3.

- **EduCore Tools for Teaching the Common Core.** This project was created and sponsored by ASCD (formerly the Association for Supervision and Curriculum Development), with funding from the Bill and Melinda Gates Foundation. The site makes available for downloading various documents that address challenges and ideas for districts and teachers in moving toward implementation of CCSSM. Videos and virtual conferences are also available from http://educore.ascd.org /channels/0e7ad8a6-1615-40be-8582-7a5ff9a3c083.

Resources to Support Access and Equity in Implementing the Common Core

One of the main principles of *Principles and Standards for School Mathematics* (NCTM 2000) is the Equity Principle, calling for high expectations and strong support for all students. The strength of any program in mathematics education can be measured by whether all students have access to a curriculum and teachers that challenge them to learn but also offer resources to help them reach their learning potential. Thus, it is not surprising that the NCTM Position Statement *Supporting the Common Core State Standards for Mathematics* (2013) highlights the need for instruction aligned with the CCSSM "to be rooted in and promote principles of access and equity" (2013). The authors of CCSSM concur and insist that the standards should not be changed but that schools should provide necessary accommodations that "allow students to learn within the framework of the Common Core" (NGA Center and CCSSO 2010). Some conditions that will help to promote access and equity in this CCSSM environment include (a) time for teachers, curriculum coordinators, and other stakeholders to adjust to the new recommendations and interpret the standards, (b) ongoing professional development of teachers so that they can effectively and appropriately engage all students in the mathematics content described in the standards, and (c) funding for research to ensure implementation and assessment to meet the intended goals of the standards and the needs of all student populations.

CCSSM and the Principles and Standards enunciated by NCTM call for high expectations for all students, yet in many school systems across the United States obstacles to equity and access persist. For example, students who have performed poorly in a traditional mathematics class are often identified as "low ability" and placed in basic mathematics classes that do not challenge them to learn. As a result, and by default, such students may fulfill the expectation that they cannot learn mathematics. Current standards documents and the principles of access and equity described above are clearly at odds with such a situation. This is similarly true for students who reside in low-income school districts where they are much less likely to experience high-quality, effective teaching in mathematics and are thus not provided with the same opportunities as students in more affluent districts. To move toward access and equity for all students, teachers, administrators, and policymakers must recognize the need for a change in beliefs and in how schools are funded. In addressing the need for a change in the beliefs of teachers and students, Dweck (2010) calls for a change to a "growth mind-set," or the belief that intelligence can be developed and nurtured in all learners—students and teachers alike—through effort, effective instruction, and differentiated supports. Such a belief is echoed in the recommendations set forth in NCTM's *Principles to Actions* (2014). In addressing the need for a change in funding to schools, we can say only that we have yet to see how many of the individual states will fund the shift to and implementation of the Common Core State Standards.

Consideration of special student populations is another component of access and equity that needs to be addressed in relation to implementing CCSSM. Special populations include students with learning disabilities, English language learners, and other underserved groups in the school community. The Common Core State Standards website does not provide explicit guidance to teachers or to states on how to work with such populations within the framework of the standards, but some of the resources mentioned previously in this chapter do offer accommodation guidelines, particularly with respect to assessment of student progress and achievement. Furthermore, the Smarter Balanced Assessment Consortium (SBAC) and the Partnership for Assessment of Readiness for College and Careers (PARCC) offer guidelines on how and when to assess special populations of students. Both assessment consortia have developed frameworks for school administrators and teachers in the consortium states for understanding available supports, tools, and accommodations for special student populations. More information about the guidelines and frameworks provided by SBAC can be found at http://www. smarterbalanced.org/parents-students/support-for-under-represented-students/. The PARCC documents are available at http://www.parcconline.org/assessments /accessibility.

Full implementation of the Common Core State Standards is being phased in or is already in place throughout most school districts across the United States. Some states and districts have already begun using the assessments developed for the Common Core and provided by either SBAC or PARCC, and achievement results are already raising questions about accountability for teachers and school administrators. However, it is clear that a great deal of time will be needed to study effects of the implementation, to make sense of assessment results, and to develop more appropriate classroom materials and teacher resources. Developing and launching a common nationwide mathematics curriculum are big steps, and implementing it with fidelity to the intended content and mathematical practices described in CCSSM is a major undertaking. At present, many teachers and districts are rushing to meet deadlines for implementation but in the process are skipping vital steps as they move from their previous curricula to CCSSM. Districts need carefully developed outlines to guide them in the transition from their present grade-level content to the grade-level content suggested by CCSSM. In addition, a plan needs to be developed for professional groups nationwide to provide input to ensure continuing renewal, or adjustments, of the CCSSM standards once schools have had the chance to work with them. Such a plan might take the shape of NCTM's efforts to adjust the NCTM Standards from the initial release in 1989 to the release of the revised and updated recommendations in 2000. How revisions and updates are made to the Common Core will be critical factors in the survival of this common national curriculum in the United States.

Chapter 7: Changes Surrounding the Teaching of Mathematics

This chapter focuses on three topics that are undergoing rapid change as this report is being written. The first topic relates to the structure and governance of schools in the United States. This includes the differences of the varied structures and what information is present on students' mathematics achievement in charter, home-schooling, and private schooling versus public education.

The second topic deals with the structure and nature of teacher education as it begins to break its bounds as a campus-based program operating through a college of education. Changes are taking place with respect to alternative processes of teacher certification and beyond, as suggested by the academies outlined in the Every Student Succeeds Act (2015). In addition, departments outside the traditional colleges of education have developed web-based programs of certification, and for-profit education entities have entered the certification business as well. The second topic in this chapter deals with these issues, describing the landscape and introducing some new features to consider.

The third topic relates to the rise of data and statistical concepts as content for the school mathematics curriculum, both for courses and for topics to permeate the entire school curriculum. Many recommendations have been made about the role of data and statistics in the mathematical sciences (including statistics) curriculum, but the focus here will be on viewing those recommendations with an eye to the much-needed statistical education for teachers of such a curriculum.

The Emergence of Charter, Home, and Private Schools

Alternate forms of schooling have been a part of North American education since before the official founding of our country. The first forms of education were religious-based schools taught by clergy accompanying groups of colonists to North American shores. As towns grew in size, private schools became more prominent, but in some cases towns provided free elementary education for primary students.

By the middle of the 1800s, most students had access to a universal, free, compulsory primary school, but numerous private academies provided a more structured curriculum and served students from the upper classes. By 1890, the dominance of private schools had ended. Public high schools had appeared on the scene and, combined with the free public elementary schools, started the shift toward free public K–12 school systems, although they enrolled less than 35% of the nation's students. A decade later, the percentage of K–12 students educated in private schools had fallen below 8%. Across the nation, a large group of students remained unaccounted for, including the rural poor and those whose racial or ethnic origins put them outside the mainstream. Many students were homeschooled—in many cases, because of parental concerns about the available schools, and in other cases, for monetary reasons. The following sections trace the models of charter schools, home schools, and private schools—sectarian and nonsectarian—in the United States today.

Public charter schools

A public charter school is a publicly funded school that is typically governed by a group or organization under a legislative charter (or contract) with the state or jurisdiction in which it exists. The charter exempts the school from certain state or local rules and regulations. In return for flexibility and autonomy, the charter school must meet the accountability standards listed in its charter. Unlike public schools, whose operation is quite similar from state to state, public charter schools operate under charters that differ greatly, as do the groups allowed to authorize the charters in states. The National Association of Charter School Authorizers, a professional organization devoted to developing a unified and professional approach to the issuance and voiding of charters based on examinations and performance of schools with respect to their charters, provides a wealth of information on public charter schools. Data collected by the association in 2013–14 found the following:

- 951 school district authorizers;
- 47 higher education institutions;
- 19 not-for-profit organizations;
- 18 state education agencies;
- 15 independent chartering boards; and
- 3 non-education governmental bodies. (National Association of Charter School Authorizers 2015 p. 3)

This list alone shows the reasons for the diversity in charters issued and charter school oversight.

The first law allowing the establishment of charter schools was passed in Minnesota in 1991. But by 2012–13, 42 states had established laws delineating the rights and responsibilities of public charter schools. Charter school legislation differs widely across the United States, reflecting regional differences in views of the purposes of public education and public financial support for public schools. California, which has the largest number of charter schools, does not allow the granting of a charter to an already existing private school, whereas, by contrast, both Arizona and Michigan allow such conversions. At the same time, both California and Michigan require charter school teachers to be certified in the state, whereas Arizona does not require teacher certification of its charter school teachers.

Data in the *2014 Digest of Education Statistics* (NCES 2015a) show that, from the 1999–2000 school year to the 2012–13 school year, the number of public schools that were charter schools had increased from 1.7% to 6.2%, a numerical growth from 1,600 to 6,100 schools. This growth entailed a significant increase in the mean number of students per school (NCES 2015b). However, this expansion has increasingly been put under fiscal and academic achievement scrutiny as some have come to perceive the growth of public charter schools as—

- draining precious public funding from already existing public schools to create new public schools that can "cherry-pick" their students in an era of increased pressure for school achievement accountability;
- creating schools that are supported by public funding but whose leadership and direction are in the hands of individuals who are not publicly elected; and
- developing a culture of elitism in public schooling that is at odds with the traditional view of schools as the mixing pot fostering the health and well-being of democracy in the United States.

What does research say about the academic achievement of charter school students? David C. Berliner and Gene V. Glass (2014) deliver a scathing review of the academic performance of charter schools. They point to the practice in some charter schools of "flunking out all but the most able" on their way to achieving high rankings in lists of effective schools without revealing the carnage involved in getting the few high achievers to the top. In 2009, the Center for Research on Education Outcomes (CREDO 2009) pointed to a finding that charter schools perform neither better nor worse than their public, non-charter school peers. The CREDO (2013) study was a repetition enlargement of sample and extension of the 2009 study. CREDO (2013) found advantages for the charter schools in mathematics and reading and especially for newer charter schools. CREDO (2013) also found that 193 of the charter schools that had been in their 2009 study were now closed. The researchers in the 2013 study then developed five scenarios that were researchable from simulations on CREDO data and showed the effect that the removal of various sectors of the sample might have on a comparison of students' academic performance in a charter versus a non-charter public school. In each case, the finding of not better and not worse still held.

A second way of examining the effects of charter schools on achievement is to analyze data from the 2015 Main NAEP Grades 4 and 8 mathematics assessments for the national public non-charter schools versus the national public charter schools with respect to mathematics. The composite NAEP scale score for grade 4 was 240 (with standard error 1.2) for the public non-charter and 236 for the public charter (with standard error 0.3). This indicates a significant difference $(p < 0.05)$ favoring higher achievement in the national public non-charter schools. Performing the same analysis on the 2015 Main NAEP Grade 8 composite mathematics assessment data, the scale score outcomes for mathematics were 281 (with standard error 0.3) for the public non-charters and 279 for the public charters (with standard error 2.1). This indicates no significant difference ($p \geq 0.05$) between the achievement levels of the charter or non-charter school format across the two forms of public schooling at grade 8 (NCES 2015c).

A third study compared charter schools' achievement to public non-charter schools' achievement on the 2003 NAEP Grade 4 mathematics test, using hierarchical linear modeling. Henry Braun, Frank Jenkins, and Wendy Grigg in their 2006 study, *A Closer Look at Charter Schools Using Hierarchical Linear Modeling*, examined the effect of being in a public, non-charter school as compared with being in a public charter school by adjusting out the effects of students' demographic backgrounds. They chose as mediating factors data available through the NAEP assessment demographic data questionnaire. They first made a gross comparison of all charter schools in the NAEP assessment with all public schools in the NAEP assessment in grade 4 mathematics. This comparison showed a 5.8-point deficit for charter schools relative to public non-charter schools. This deficit adjusted down to 4.7 points when they adjusted the total variance of NAEP mathematics scores into the portion attributable to differences among students within schools and the portion attributable to differences among schools.

Braun, Jenkins, and Grigg (2006) tested four models—(1) school type, (2) adjusting for race and then for school type, (3) adjusting for race and other student characteristics and then adjusting for school type, and (4) adjusting for race and other school characteristics and then adjusting for school type and other student characteristics. The school types were charters affiliated with public school districts and charters not affiliated with public school districts. For charters affiliated with public school districts, all the models showed no significant difference between the non-charter public schools and the charter schools affiliated with a public school district. However, the estimates for the non-public school-affiliated charters were negative for all four models and statistically significant. When the third model (adjusting for race and other student characteristics and then adjusting for school types) was applied, the estimated difference was −6.4 points, which was smaller than the estimated school type contrast of −9.8 when there was no adjustment at level 1. The effect size of this contrast was 0.23. These results suggest that under any of the models, charter schools not affiliated with school districts achieve at a significantly lower level than public schools, even with the adjustments described in the 2003 Grade 4 NAEP Mathematics assessment.

Homeschooling

Homeschooling has been with us since the days when the first Europeans set foot on North America and undoubtedly occurred in the families of the indigenous people already living here. One of the reasons driving many parents to elect to homeschool their children is that they believe that they have a better sense of what knowledge their children need than the public schools do. Others fear the potential "contamination" of their children by ideas brought to school by other children, or even their children's teacher, at a public school. Other reasons connect with geographical locations relative to public schools for people in very rural areas, and religious reasons for others who wish to keep their children isolated from the influences of other religions during their developmental years (Jeub 1994).

But homeschooling is not just a simple issue of keeping children at home and teaching them some basic skills, as some would like to make it out to be. The Internet is filled with ideas, resources, and research on homeschooling. The main argument advanced in favor of homeschooling is that homeschooled students come to high school or college when they leave the home classroom more socialized to learn and interact with adults than their new non-homeschooled peers are. Further, their test scores on standardized assessments are in the range of 10% to 25% higher than those of their average peer, mainly as a result of having to figure out a lot of things for themselves (Ray 2000, 2015).

States require homeschooling parents to satisfy a few state laws regarding the education of children within the state. These requirements may include the following:

- Complying with the state's compulsory school attendance age
- Formally withdrawing a child from his or her current school
- Complying with the state's homeschool law—teaching the required subjects, providing instruction in English, and knowing when and how to legally refer to your school
- Keeping the required state records on your child's education—attendance, texts, samples of work (this is especially important during the high school years for use during college admission, etc.). (Home School Legal Defense Association [HSLDA] 2015)

States vary in their regulations for those conducting a home school. The rules are much the same for individual parents who band together to develop a group home school. This in essence becomes more like a private school and usually entails complying with more regulations.

One of the issues that arose with the No Child Left Behind Act was the issue of mandated testing (U.S. Department of Education 2008). The Common Core testing programs or equivalent programs spawned by the Race to the Top legislation and modifications in the NCLB law resulted in mistaken attempts to test homeschooled children. This is one example of a case in which intentions and misunderstandings led to tensions between those supporting public schooling and those supporting homeschooling (HSLDA 2015).

The absence of homeschooled students from state, NAEP, and other educational research samples of the nation's youth is an interesting conundrum. Nonparticipation is a personal right as viewed by the homeschooling portion of the population, but at the same time the absence of these individual students from the nation's pool of students creates a misunderstanding of what U.S. students know and can do in mathematics and other disciplines. It also creates misapprehensions about private and public schooling on both sides of the divide between backing parents' rights to homeschool and considering the concerns of the remaining members of the public about the possible negative effects of this removal of homeschooled students from interactions with the public school. Christopher Lubienski (2000) has examined the divide and the common good in an article contained in a special double issue of the *Peabody Journal of Education* in 2000. William Schmidt (2012), a researcher best known for his activity in international comparative studies, including TIMSS and PISA, echoes some of the same concerns about the fractionating of society by segregating various subgroups from the total public in terms of opportunity and contributions of all to the common good. Brian Ray, a researcher at the National Home Education Research Institute and coeditor of the special issue of the *Peabody Journal of Education* on homeschooling, provided perhaps the best overview of studies examining the achievement outcomes of homeschooled youth on standardized assessments (Ray 2015).

Ray's review provides two types of information. First, it offers a careful listing of the major studies that have reviewed homeschooled youths' experience on standardized and state-level achievement tests. Second, it provides a look at a study attempting to establish some causal factors explaining homeschooled youths' performance on such instruments. The review of literature suggests that homeschooled students perform, on the state tests, Stanford batteries, and Iowa test, somewhere between the 67th and 87th percentiles of U.S. students, depending on which batteries are involved and the ages of the students. The clear takeaway is that homeschooled students, as a group, are doing as well as or significantly better than the U.S. non-homeschooled youth on standardized testing instruments (Ray 2000). It is also interesting to note that the home-based factors cited in Barton and Cooley's analysis of early school success are also factors that are high on the list of homeschooling characteristics (2007).

Private schools

The final grouping of non-public schools is private schools. Students attending these schools do so because of parental choice, as do students in the two previous categories (Latham 1998). Private schools can be separated into those established along doctrinal lines within religious denominations and those established by groups of parents wishing to provide an alternative to either religious private schools or the existing public schools in their geographical area. Some of the more famous college preparatory private schools are boarding schools. The curricular differences of these schools from public schools include the addition of religion to the curriculum in religion-linked private schools and the accelerated curriculum in college preparatory private schools. Although these private schools are not always required to have certified teachers, they do—especially the secular ones—

strive to have highly qualified teachers, because such teachers are a draw in the recruitment of students. A recent government survey of state requirements for private schools indicates that 28 out of 50 states have some sort of teacher certification requirements for at least some state-specified private schools. Further, 39 out of 50 states have curriculum requirements for at least specified schools, and 48 of 50 states require state-specified private schools to complete a reporting process with the state in which the school is located (U.S. Department of Education 2009).

The size and magnitude of the various educational structures in the U.S. delivery of education are noted in Chapter 1. The NAEP assessments provide breakdowns in student attainment of the NAEP goals for public and private schools. In addition, in this chapter, we have reported other research findings that add what is known about homeschooling. The most recent NAEP achievement results provide data for public, charter, and various forms of private schooling for grade levels by scale scores and achievement levels.

Table 17 contains the data from the public non-charter versus the public charter comparisons discussed earlier, as well as from the public versus Roman Catholic comparisons (note that charter schools are already in the public sample). At grades 4 and 8, the choices for the question on charter schools did not permit responses on private schools, Bureau of Indian Affairs schools, or DoDEA schools. At grade 8, the question allowed responses to include other private schools and both Bureau of Indian Affairs and DoDEA schools, but these unfortunately did not draw sufficient sample size to meet requirements for analysis of differences among all the various categories of schooling.

At grade 4, the results in table 17 show a significant difference ($p < 0.05$) for higher mathematics achievement in 2015 in the non-charter public schools over the charter public schools. In the public versus private comparisons of performance at grade 4, the national public fourth-grade students had significantly lower mathematics achievement ($p < 0.05$) than their counterparts in the Roman Catholic private schools. However, the sample sizes were insufficient to make other private school comparisons.

Table 17

Analysis of 2015 NAEP grade 4 mathematics results by public, charter, and private school NAEP scale score and performance level

Group	Mean scale score	S.E.	Percent Proficient or above	Group	Mean scale score	S.E.	Percent Proficient or above
National public non-charter*	240	(0.3)	40	National public**	240	(0.3)	39
National charter*	236	(1.2)	35	Catholic**	247	(1.2)	48
Other schools*	†	†	†	Other schools **	†	†	†

* National public non-charter includes all public non-charter. National charter includes all public charter. The category "Other schools" contains private schools.

**National public includes all public (charter and non-charter) schools. Catholic contains Roman Catholic schools. The category "Other schools" contains other private (religious and secular, Bureau of Indian Affairs schools, and DoDEA schools).

† Missing data due to question wording or insufficient sample size to answer.

Table 18 shows corresponding results at grade 8. These results show a significant difference favoring the public non-charter group, which demonstrated higher achievement on the 2015 NAEP grade 8 mathematics test when compared with the public charter group ($p < 0.05$). The comparisons for the public versus private school performance at grade 8 were limited to public schools compared with Roman Catholic schools because of insufficient sample sizes for the other choices of private schooling. The comparison of public grade 8 students' mathematics achievement with that of their Roman Catholic private school counterparts significantly favored higher achievement among the eighth graders in Roman Catholic private schools ($p < 0.05$) (NCES 2015c).

Table 18

Analysis of 2015 NAEP grade 8 mathematics results by public, charter, and private school NAEP scale score and performance level

Group	Mean scale score	S.E.	Percent Proficient or above	Group	Mean scale score	S.E.	Percent Proficient or above
National public non-charter*	282	(0.3)	33	National public**	281	(0.3)	32
National charter*	279	(2.1)	30	Catholic**	293	(1.5)	44
Other schools*	†	†	†	Other private**	†	†	†

* National public non-charter includes all public non-charter schools. National charter includes all public charter. The category "Other schools" contains private schools.

**National public includes all public (charter and non-charter) schools. Catholic contains Roman Catholic schools. The category "Other schools" contains other private schools (religious and secular, Bureau of Indian Affairs schools, and DoDEA schools).

† Missing data due to question wording or insufficient sample size to answer.

A final study comparing public and private schools was that described by Christopher and Sarah Lubienski in their 2013 book, *The Public School Advantage: Why Public Schools Outperform Private Schools.* Using approaches that paralleled analyses carried out by Braun, Jenkins, and Grigg (2006) in their comparison of public and charter schools, the Lubienskis built predictor variable scores by segmenting and merging data from the NAEP demographic information with other available demographic information to adjust the NAEP assessment scores. Although one can perhaps take exception to the statistical approaches, the fact remains that their careful statistical examinations of the data returned results favoring public schools over private schools. While supporters for the private school model remain, they have, to date, not presented as convincing a set of studies backing their choice of schools.

The comparison of school formats and achievement produced without a discussion of other factors such as home environment, educational achievements of students' parents, general socioeconomic factors of the community in which the school resides, and a myriad of other considerations leaves such comparisons open to question. Perhaps the most important factor is the background and skills honed by experience of the students' teachers. The next portion of this chapter turns to changes in teacher education that are already under way and some that have been proposed but have not yet been implemented.

Teacher Education and Its Continuing Role in the Education of Teachers

Teacher education requirements in the United States have not changed much over the past several decades. A bachelor's degree and a teaching certificate are needed to teach in most public schools at any level, kindergarten through secondary education. The teaching certificate is generally obtained through satisfactory completion of a combination of courses taken at the college level and in-school experiences, such as observations and supervised practice teaching, in or around the grade levels at which the teaching certificate is requested. Most states also require that the teacher candidate pass one or more tests, which usually assess specific subject-matter knowledge and general knowledge about teaching and the education system.

All states, with the exception of Alaska and Oregon, also provide some alternative route to teacher certification based on an individual's prior experiences, education, and, potentially, a bundled set of courses and internship experiences (National Center for Education Information [NCEI] 2010). NCEI estimates, based on data submitted by the states, indicate that at least 59,000 individuals were issued certificates to teach through alternative routes in 2008–9. There does not appear to be a public document providing equivalent data on the numbers of alternately certified teachers on an annual basis since this figure. Also, in cases of teacher shortage or the movement of a teacher from one state to another, provisional certification is possible through state education officials until the relocated teacher has met all the requirements for full certification. A new provision, passed as part of the Every Student Succeeds Act (ESSA) in December 2015, includes funding for the development of what are defined as teacher, principal, or other school leader preparation academies in Title II—Preparing, Training, and Recruiting High-Quality Teachers, Principals, or Other School Leaders. Such academies may be established by public or nonprofit entities for the purpose of preparing teachers, principals, or other school leaders. Using a model similar to that for authorizing charter schools, the governor of each state may designate an entity to authorize such academies. Students completing an academic year of evidence-based coursework and integrated clinical experiences will receive a certificate of completion. This certificate is to be, as determined by the state, "at least the equivalent of a master's degree in education" for purposes of hiring, retention, compensation, and promotion." This creation of a new track to the classroom has raised concerns in the education community about the incursion of nonprofit entities, similar to the "degree mills" that have received unfavorable publicity in recent years, into the teacher-preparation world with governmental approval (ESSA 2015)

For teacher preparation in mathematics, university mathematics departments typically offer the mathematics courses taken by preservice and in-service teachers as part of their training, although in some institutions' education departments may offer some of those courses intended for preservice elementary school teachers. In these institutions, the methods courses for the teaching of mathematics may be taught in either the department of mathematics (if the mathematics educators are housed there), or the college of education.

In 2005 and again in 2010, the Conference Board of the Mathematical Sciences (CBMS) survey of undergraduate mathematics programs asked a special set of questions focused on coursework offered for prospective teachers of kindergarten through grade 12 (Blair, Kirkman, and Maxwell, 2013; Lutzer 2007). For universities at which teacher-preparation programs exist, the questions focused on where such programs were housed and what courses and experiences were required of the students in these programs. The results of the 2015 survey are not yet available; however, results from the 2010 survey indicated that 72% of the institutions had a K–8 teacher certification program. This was a decline from 84% in 2000 and 87% in 2005 (Blair, Kirkman, and Maxwell 2012). The reasons for the decline in the percentage of institutions offering such a program were not clear in an era of recommendations calling for more mathematics specialists in the K–8 years and movement in states to provide special certification of teachers at these levels. An examination of some subareas within the data groups indicate that the major sources of the decline were at universities with PhD programs in mathematics and four-year colleges offering the BA as their highest degree in mathematics. A slight increase was seen in the percentage of K–8 teacher-preparation programs at universities offering the MA in mathematics as their highest degree. This perhaps may indicate that these specialist programs are most often found at the universities that have grown out of what historically were known as "normal schools," which were teacher-training colleges. The CBMS 2015 survey results may indicate positive changes, since several states are beginning to put in place "specialist" programs for K–8 mathematics teachers that lead to advanced degrees.

Examination of the data for four-year colleges that provide programs of preparation for "early grades mathematics specialization" reflect that, on average, 42% of such programs require two mathematics courses, 14% require three mathematics courses, 14% require four mathematics courses, and 11% require five or more mathematics courses. Across all of these programs, the average course requirements are 2.7 mathematics content courses and 1.4 mathematics pedagogy courses, taught either within or outside the mathematics department, and 0.5 courses in mathematics pedagogy, taught within the mathematics department (Blair, Kirkman, and Maxwell 2012).

With respect to the full range of content and pedagogy courses offered to students by mathematics departments providing preparatory programs for preservice K–grade 8 teachers, the data show that 74% of these departments offer courses in number and operations, 57% offer courses in algebra, 69% offer courses in geometry/measurement, 56% offer courses in statistics or probability, and 31% offer courses dealing with teaching methods of elementary mathematics (Blair, Kirkman, and Maxwell 2012).

CBMS's 2012 report *The Mathematical Education of Teachers II* (MET II) and ASA's 2015 document *The Statistical Education of Teachers* (SET; Franklin et al. 2015) provide comprehensive frameworks for developing or modifying teacher-preparation programs and include suggestions for continuing professional development as well. These documents offer significant guidance regarding the nature of courses that would be valuable to teachers to build a strong understanding of mathematics and statistics for teaching. Both documents stress that such coursework should not focus on traditional mathematics content but instead should encompass content that considers the mathematics that teachers need to know and how they need to know it to be effective in the classroom. Unfortunately, the CBMS's 2015 survey of undergraduate mathematics programs did not include focused questions about the mathematics course offerings for students in secondary mathematics teacher preparation programs. Perhaps other elements of the survey will document significant changes to teacher preparation in mathematics that align with the recommendations of the SET and MET II documents (Franklin et al. 2015; CBMS 2001, 2012).

Standards for and other changes to mathematics teacher preparation

A few current trends in teacher preparation are worth mentioning here since they affect teacher preparation in mathematics as well. One of these is an increased interest in training teachers in science, technology, engineering, and mathematics (STEM) subjects. Reflecting that trend, the UTeach Program was developed at the University of Texas at Austin in 1997 with the goal of increasing the number of qualified STEM teachers in

U.S. secondary schools. The program has since been replicated at as many as 39 universities and four-year colleges nationwide. The program's hallmark is its requirement that students begin teacher preparation coursework and field experiences during their first year in the program, thus allowing them time for a concentrated focus on a STEM major while pursuing secondary teaching certification during a typical four-year plan.

According to the program's website, "Faculty with expertise in STEM fields, in STEM teaching and learning, and in the history of science and mathematics worked alongside master teacher practitioners (UTeach clinical faculty) to design a program emphasizing deep understanding of STEM content, practices and pedagogy, and strong connections between theory and practice" (UTeach Institute 2015). The UTeach Institute supports other teacher preparation programs that wish to replicate this recruitment and training program and also allows for continuous improvement of the UTeach model.

Another recent trend in teacher preparation is the move toward standardized performance assessments of teacher candidates. Teacher Performance Assessments (TPAs) are based on standards recognized by teacher educators as important knowledge and skills needed by novice teachers. Such assessments go beyond subject-area tests to requiring teacher candidates to demonstrate their ability to carry out lessons in specific content areas that have a positive impact on student learning. The assessments are intended to support preservice teacher growth and also serve as an assessment system for teacher-preparation programs (Darling-Hammond and Hyler 2013).

One such nationally available TPA program in the United States is the edTPA. This assessment and support program requires aspiring teachers to demonstrate readiness to teach through lesson plans designed to support their students' strengths and needs, to engage real students in ambitious learning, to analyze whether their students are learning, and to adjust their instruction to become more effective The edTPA builds on decades of teacher performance assessment development and research regarding teaching skills and practices that improve student learning (edTPA 2015).

This assessment method has the potential to create a more rigorous teacher preparation process that could improve teaching and have a positive impact on student achievement (Mehta and Doctor 2013). However, some have already voiced opposition to the edTPA program because they see it as another way for a testing company, Pearson, to make money and another set of hoops for prospective teachers to jump through to become certified (Madeloni and Hoogstraten 2013). The effectiveness or lack of effectiveness of this assessment method has not yet been established through evidence.

Teacher workforce

Data on the quantity and quality of the teacher workforce come from education agencies in individual states, the National Survey of Science and Mathematics Education (Banilower et al. 2013), a survey on schools and staffing conducted by the National Center for Education Statistics (NCES), and the National Assessment of Education Program (NAEP) assessments. The state data are not always complete, and some of the data raise questions about accuracy as well as completeness. Data from the most recent NAEP assessments have only recently been released, and results are reported elsewhere in this document. Thus, the remainder of this section focuses on the results of the 2012 National Survey of Science and Mathematics Education in relation to the status and background of teachers of mathematics in elementary school (Malzahn 2013), middle school (Fulkerson 2013), and high school (Smith 2013) and the 2012 Schools and Staffing Survey (SASS) on the qualifications of public high school–level teachers (Hill, Stearns, and Owens 2015).

Data from the *Education and Certification Qualifications of Departmentalized Public High School–Level Teachers of Selected Subjects: Evidence from the 2011–12 Schools and Staffing Survey* (Hill, Stearns, and Owens 2015) indicate that mathematics students in public high schools in 2011–12 were disproportionately more likely to have a teacher with neither an undergraduate major in mathematics nor certification in mathematics (10.4% of students), as compared with students in English (8.0% of students), science (7.0% of students), or the social sciences (5.7% of students). Similarly, the percentage of teachers who reported mathematics as their main assignment and had also majored in mathematics was 70.1% in 2011–12, as compared with 79.4% of English teachers majoring in English, 74.4% of biology teachers majoring in biology, and 78.9% of social studies teachers majoring in a social science (Hill, Stearns, and Owens 2015).

The report *National Survey of Science and Mathematics Education: Status of Elementary School Mathematics* (NSSME; Malzahn 2013) indicates that only 4% of elementary teachers who teach mathematics in K–grade 5 have an undergraduate degree in mathematics or mathematics education. Even so, 95% of elementary school teachers report taking at least one post-baccalaureate class in mathematics content for elementary school teachers, and 77% of K–grade 5 teachers who teach mathematics perceive that they are well prepared to teach mathematics to their students. At the middle school level, roughly 36% of mathematics teachers earned an undergraduate degree in mathematics or mathematics education, and the vast majority of middle school mathematics teachers completed at least one course in mathematics education (88%), although coursework in other mathematics subjects ranged widely (Fulkerson 2013). Around 65% of middle school teachers enrolled in statistics or calculus in college, although only 21% took a college course in Euclidean or axiomatic geometry.

At the high school level, the data demonstrate stronger mathematics backgrounds for mathematics teachers. Seventy-five percent of high school mathematics teachers have earned undergraduate degrees in mathematics or mathematics education (Smith 2013). As with middle school mathematics teachers, almost all have taken a course in mathematics education (90%) or have completed student teaching in mathematics (82%). Most took college courses in a wide range of mathematics areas, such as calculus (95%), statistics (86%), linear algebra (84%), and computer science (79%). Fewer have had coursework in geometry (59%) or discrete mathematics (54%). Eighty-four percent of high school teachers perceive that they are well prepared to teach mathematics in grades 9–12.

Another important component of the effectiveness of mathematics teachers is their commitment to their own growth as teachers. The NSSME report demonstrates that the majority of mathematics teachers have taken advantage of opportunities for professional development in mathematics teaching within the past three years (87% of teachers at the elementary school level and 89% at the middle and high school levels).

Another area of interest is the amount of knowledge and coursework that teachers have in general in the area of data and statistics. Statistics is a discipline that is rapidly growing in importance as an aid in decision making across the board in daily applications at home as well as in work and career applications. The next section addresses issues in educating teachers to teach statistics and reviews recommendations made in the SET report mentioned earlier (Franklin et al. 2015).

Educating Teachers to Teach Statistics

Statistical reasoning skills have become critical to individuals' functioning in modern society. These skills are in high demand in business, industry, government, and academic settings, and the shortage of people with them is only expected to grow over the next decade. To address this need, CCSSM and other standards for pre-K–grade 12 mathematics place much more emphasis on statistics and probability than have previous curricular recommendations. The training of teachers to offer instruction aligned with these new standards needs to be updated to accommodate this increased emphasis on statistics in pre-K–12 education.

The American Statistical Association (ASA) commissioned a report, *The Statistical Education of Teachers* (SET; Franklin et al. 2015), to clarify the recommendations on the preparation of teachers to teach statistics found in the report *Mathematical Education of Teachers II* (MET) from the Conference Board of the Mathematical Sciences (CBMS 2012). Additionally, SET addresses the professional development of currently practicing teachers, highlights the differences between statistics and mathematics that have important implications for teaching and learning, illustrates the statistical problem-solving process across all levels of development, and makes pedagogical recommendations relevant to teaching statistics at all levels. SET builds on a previous report from ASA, *Guidelines for Assessment and Instruction in Statistics Education (GAISE Report)* (Franklin et al. 2007; http://www.amstat.org/education/gaise/).

The SET report (available at http://www.amstat.org/education/SET/SET.pdf) identifies the features of statistics that differ from mathematics and discusses the implications of these differences on the preparation of teachers to teach statistics. In particular, statistics differs from mathematics in its focus on variability and its greater emphasis on context in problem solving. Mathematics focuses on deterministic questions in which variability does not play a role and typically places less emphasis on problem context. In statistical reasoning, data and context are inextricably linked, and measuring variability within a given context is an important component of data analysis. SET strongly recommends that the emphasis in statistics teacher education be on statistical thinking and include training in applied statistics and the pedagogy needed to teach these concepts and processes. The consensus among statistics educators is that statistical concepts can best be taught through the use of real data in an active learning environment and that teachers need to learn statistics in this way so that they are prepared to teach in the same way.

Probability plays an important role in both mathematics and statistics, but its role differs in the two fields. Probability has an essential role in statistical reasoning, and an understanding of probability is critical for interpreting confidence intervals and significance tests. Additionally, probability provides models for quantifying variability and serves as the basis for random assignment and random sampling. In mathematics, probability is in itself an important subfield of study that includes topics less relevant to statistical practice. The SET report recommends that probability concepts be developed through simulation and the use of theoretical distributions such as the normal distribution, but axiomatic approaches to probability are not recommended. The report strongly recommends that probability modeling be included in teacher preparation programs.

Pedagogical training of statistics teachers must also introduce the role and use of technology and assessment in statistics teaching. Technology should be used to develop concepts and to analyze data. The use of simulations can greatly enhance concept development, and many resources are available on the Internet for both teaching and learning through simulation (http://www.amstat.org/education/usefulsitesforteachers.cfm).

Assessments in statistics should emphasize conceptual understanding rather than merely mechanical skills. SET points out that for more than 20 years, statistics educators have called for a greater emphasis on statistical concepts in all assessment of student learning. Unfortunately, at present, large-scale national assessments still predominantly assess procedural competency in statistics, and this presents a conflict with the preferred pedagogy because teachers are evaluated on the basis of students' performances on these tests. Two projects that align tests with conceptual learning of statistics are the Assessment Resource Tools for Improving Statistical Thinking (ARTIST) project (https://apps3.cehd.umn.edu/artist/index.html) and the Levels of Conceptual Understanding in Statistics (LOCUS) project (https://locus.statisticseducation.org/). The goal of the ARTIST project, developed by statisticians from the University of Minnesota and California Polytechnic State University, is to "to help teachers assess statistical literacy, statistical reasoning, and statistical thinking of students in first courses of statistics." Although primarily intended for college students, assessment items found at the ARTIST website are appropriate for high school students and are freely available to teachers. The LOCUS project, led by statistics educators at the University of Florida, University of Minnesota, and the Educational Testing Service and supported by funding from the National Science Foundation, has developed items and instruments that assess conceptual understanding of statistics as articulated in the *GAISE Report*. The LOCUS assessments are appropriate for students in grades 6–12 and can also be used to assess understanding of preservice and in-service teachers.

The SET report points out that statistics and mathematics share common ground, especially in the eight Standards for Mathematical Practice as outlined in CCSSM. SET examines these eight standards from the perspective of statistics education and shows how each standard can be connected to the practices necessary to acquire and apply statistical knowledge.

The SET report does not make specific course recommendations for teachers at each level of teacher licensure—elementary, middle, or high school—but discusses the key concepts that teachers should know and be able to teach at each level. Ultimately, SET makes six recommendations for the statistical education of teachers:

> **Recommendation 1.** Prospective teachers need to learn statistics in ways that enable them to develop a deep conceptual understanding of the statistics that they will teach.
>
> **Recommendation 2.** Prospective teachers should engage in the statistical problem-solving process—formulate statistical questions, collect data, analyze data, and interpret results—regularly in their courses.
>
> **Recommendation 3.** Because many currently practicing teachers did not have an opportunity to learn statistics during their preservice preparation programs, robust professional development opportunities need to be developed for advancing in-service teachers' understanding of statistics.
>
> **Recommendation 4.** All courses and professional development experiences for statistics teachers should allow them to develop the habits of mind of statistical thinkers and problem solvers, such as reasoning, explaining, modeling, seeing structure, and generalizing.
>
> **Recommendation 5.** At institutions that prepare teachers or offer professional development, statistics teacher education must be recognized as an important part

of a department's mission and should be undertaken in collaboration with faculty from statistics education, mathematics education, statistics, and mathematics.

Recommendation 6. Statisticians should recognize the need for improving statistics teaching at all levels. Mathematics education, including the statistical education of teachers, can be greatly strengthened by the growth of a statistics education community that includes statisticians as one of many constituencies committed to working together to improve statistics instruction at all levels and to raise professional standards in teaching.

If the mathematical and statistical education of teachers—both preservice and in-service—are to meet the goals set by the MET II and SET recommendations, the recruitment, retention, and continual professional development of teachers of these subjects have to become a core responsibility of both the educators of teachers and the professional mathematical and statistical faculty in colleges and universities across the United States. Further, the administrations of these colleges and universities, the NSF, and U.S. Department of Education, as well as other foundations and corporations who employ people in the quantitative and decision sciences have to step up and ensure that these programs are both available and funded, and that the faculty and public school teachers who participate in them receive appropriate recognition.

Chapter 8: Programs for Special Populations of Students

High school students who complete the standard college-bound curriculum that enables them to take precalculus before entering twelfth grade have three potential paths if they want to continue the study of mathematics in twelfth grade. First, if a student's school is very small and no two- or four-year college or university is nearby, then the student may be able to take an individualized course under a teacher's guidance or over the Internet. Second, if the student's school is near a college or university, the student may be able to take a college course and apply the credit toward high school graduation. Third, if enough students in the school are in the same position as the student, then the school may wish to offer Advanced Placement (AP) courses.

Advanced Placement Programs

In 1955, under the auspices of the College Board, the Educational Testing Service (ETS) created the Advanced Placement Program to enable students to take college-level work before graduating from high school (Handwerk et al. 2008). High schools participating in this program offer courses with syllabi designed to align with introductory college courses. In the 2014–15 school year, 36 AP courses were offered in seven different disciplinary areas, with more than 21,500 high schools worldwide participating and more than 2,483,452 individual students taking at least one examination (College Board 2015b). Scores on AP tests range from a top score of 5 down to 1. The American Council on Education and the College Board have developed a matrix for universities to use in developing a policy to provide advanced placement or award credit for AP coursework (College Board 2015d). For example, the College Board has suggested that scores of 5 down to 3 correlate to levels of preparation and achievement in related universities courses in the following manner: 5 is equivalent to extremely well qualified; 4 is equivalent to well qualified; and 3 is equivalent to qualified (College Board 2015b). Lower scores are not suggested as constituting a basis for earned credit for advanced placement.

In the 2014–2015 school year, a total of 421,239 students took either the Calculus AB or Calculus BC examination. A total of 268,316 out of this total group of students (63.7%) scored a 3 or better on the individual Calculus AB or Calculus BC examination that they took. The percentages of 3 or better for the individual examinations were 57.4% for Calculus AB and 79.8% for Calculus BC (College Board 2015a, 2015g).

Most AP courses are a year in length. Many high schools, however, offer "block schedules" with longer class periods each day, and in these schedules, AP courses in some schools are compressed into one-semester configurations, an approach that both the College Board and many universities believe is extremely disadvantageous to students. In May of each year, ETS administers nationwide exams for each of the AP calculus courses. Colleges have the option of offering college credit, placing students in more advanced classes (with or without credit), or ignoring the scores that students receive. Many colleges take scores on AP tests into account when placing students into courses.

When AP courses are taken in the eleventh grade or earlier, they can be considered along with a student's application to a college and may factor into admissions decisions. Although scores on AP tests taken in the twelfth grade are not available to colleges before admissions decisions are made, enrollment itself in AP courses tends to signify that an applicant is a serious student, and if the high school is known to be scholastically oriented, enrollment in an AP course can increase the student's chances of admission to some colleges.

Advanced Placement programs in calculus

Two AP exams are offered in calculus: Calculus AB (since 1956) and Calculus BC (since 1969) (College Board 2014a). Calculus BC is not an enhancement of the topics addressed in Calculus AB, inasmuch as the two courses treat common topics requiring a similar depth of understanding. Instead, Calculus BC is an extension of Calculus AB with extra topics—most notably, L'Hospital's rule, the study of series, and more content showing the applications of calculus in problem solving. Each of the exams is scheduled for three hours and 15 minutes at the end of the year. Further, students enrolled in Calculus BC receive a Calculus AB subscore for their performance on the items on the Calculus BC examination that examine Calculus AB content. This portion composes about 60% percent of the BC examination. This subscore gives colleges and universities, as well as teachers and counselors at the students' secondary school, additional information about their performance on Calculus AB topics while students in Calculus BC courses consider more topics in the same amount of time.

The multiple-choice portion of each course examination consists of 45 questions to be completed in 105 minutes. These questions are divided into two parts, part A and part B. Part A consists of 28 items to be completed in 55 minutes and does not allow the use of a calculator. Part B, the second major part of the multiple-choice items, consists of 17 questions to be completed in 50 minutes and includes some questions for which a graphing calculator is required. Each examination also includes a constructed-response section, which consists of 6 problems to be completed in 90 minutes and is also divides into parts A and B. Part A consists of 2 problems to be completed in 30 minutes and requires the use of a graphing calculator. Part B consists of 4 problems to be completed in 60 minutes and does not allow the use of a calculator (College Board 2014a).

The syllabi for Calculus AB and BC are developed, and modified periodically, by a national committee of the College Board consisting of both secondary and university calculus teachers. In 2014, the AP Calculus Curriculum Framework was redesigned to bring a sharper focus to the "big ideas" that shape the course and provide a conceptual framework for students to connect and apply the central knowledge related to these big ideas. Calculus AB focuses on three big ideas: limits, derivatives, and integrals and the fundamental theorem of calculus. Calculus BC focuses on four big ideas, the three from Calculus AB and series.

The syllabi for Calculus AB and BC were rewritten in 2014 in a format that ties the big ideas to mathematical practices associated with students' development of the depth of understanding desired for these big ideas. The mathematical practices for AP Calculus (MPACs) are six in number. These MPACs define tools that allow a successful learner to creatively develop understanding of the big ideas while relating them to other broad sets of cognitive skills that allow such students to apply their knowledge within mathematics and in other disciplines to solving problems and advancing knowledge (College Board 2014a).

The six MPACs for AP Calculus are as follows:

MPAC 1: Reasoning with definitions and theorems

Students can—

- use definitions and theorems to build arguments, to justify conclusions or answers, and to prove results;
- confirm that hypotheses have been satisfied in order to apply the conclusion of a theorem;
- apply definitions and theorems in the process of solving a problem;
- interpret quantifiers in definitions and theorems (e.g., "for all," "there exists");
- develop conjectures based on exploration with technology; and
- produce examples and counterexamples to clarify understanding of definitions, to investigate whether converses of theorems are true or false, or to test conjectures.

MPAC 2: Connecting concepts

Students can—

- relate the concept of a limit to all aspects of calculus;
- use the connections between concepts (e.g., rate of change and accumulation) or processes (e.g., differentiation and its inverse process, antidifferentiation) to solve problems;
- connect concepts to their visual representations with and without technology; and
- identify a common underlying structure in problems involving different contextual situations.

MPAC 3: Implementing algebraic/computational processes

Student can—

- select appropriate mathematical strategies;
- sequence algebraic/computational procedures logically;
- complete algebraic/computational processes correctly;
- apply technology strategically to solve problems;
- attend to precision graphically, numerically, analytically, and verbally and specify units of measure; and
- connect the results of algebraic/computational processes to the question asked.

MPAC 4: Connecting multiple representations

Students can—

- associate tables, graphs, and symbolic representations of functions;
- develop concepts using graphical, symbolical, or numerical representations with and without technology;
- identify how mathematical characteristics of functions are related in different representations;
- extract and interpret mathematical content from any presentation of a function (e.g., utilize information from a table of values);

- construct one representational form from another (e.g., a table from a graph or a graph from given information); and
- consider multiple representations of a function to select or construct a useful representation for solving a problem.

MPAC 5: Building notational fluency

Students can—
- know and use a variety of notations (e.g., $f'(x)$, y', $\dfrac{dy}{dx}$);

- connect notation to definitions (e.g., relating the notation for the definite integral to that of a limit of a Riemann sum);
- connect notation to different representations (graphical, numerical, analytical, and verbal); and
- assign meaning to notation, accurately interpreting the notation in a given problem and across different contexts.

MPAC 6: Communicating

Students can—
- clearly present methods, reasoning, justifications, and conclusions;
- use accurate and precise language and notation;
- explain the meaning of expressions, notation, and results in terms of a context (including units);
- explain the connections among concepts;
- critically interpret and accurately report information provided by technology; and
- analyze, evaluate, and compare the reasoning of others. (College Board 2014a, pp. 4–6)

These MPACs are further used to link the big ideas to enduring understandings within each big idea, learning objectives specific to the course, and essential knowledge related to the facts, concepts, and principles central to knowing and being able to creatively and productively use the knowledge of calculus that is key to each AP course. The College Board (2014a) delineates the full structure linking the MPACs to the enduring understandings, learning objectives, and essential knowledge.

At the time of this writing, four of the most frequently used textbooks for the AP Calculus courses are the following (ordered alphabetically by author; full citations appear in the Bibliography):

- *Calculus: Graphical, Numerical, Algebraic*, 5th ed. (Finney et al., 2012)
- *Calculus: Single Variable*, 6th ed. (Hughes-Hallet et al., 2012)
- *Calculus: Early Transcendental Functions*, 6th ed. (Larson and Edwards, 2015)
- *Calculus*, 8th ed. (Stewart 2016)

Advanced Placement program in statistics

A single Advanced Placement exam is offered in statistics. AP Statistics is meant to be equivalent to a one-semester, introductory, noncalculus-based college course in statistics.

Graphing calculators with statistical capabilities are required for the exam, but the College Board emphasizes that they are not equivalent to computers in the teaching of statistics. In May 2015, 195,526 students took the AP Statistics exam, and 56.8% of these scored 3 or higher (College Board 2015c).

An outline of the areas covered by the AP Statistics examination follows (College Board 2010):

I. Exploring Data: Describing Patterns and Departures from Patterns
- Constructing and interpreting graphical displays of distributions of univariate data (dot plot, stem plot, histogram, cumulative frequency plot)
- Summarizing distributions of univariate data
- Comparing distributions of univariate data (dot plots, back-to-back stem plots, parallel box plots)
- Exploring bivariate data
- Exploring categorical data

II. Sampling and Experimentation: Planning and Conducting a Study
- Overview of methods of data collection
- Planning and conducting surveys
- Planning and conducting experiments
- Generalizability of results and types of conclusions that can be drawn from observational studies, experiments, and surveys

III. Anticipating Patterns: Exploring Random Phenomena Using Probability & Simulation
- Probability
- Combining independent random variables
- Normal distribution
- Sampling distributions

IV. Statistical Inference: Estimating Population Parameters and Testing
- Hypotheses
- Estimation (point estimators and intervals)
- Tests of significance

Additional information about AP Statistics can be found in the Curricular Framework for AP Statistics (College Board 2010).

At the time of this writing, four of the most frequently used textbooks for the AP Statistics course are the following (ordered alphabetically by author; full citations appear in the Bibliography):

- *Statistics: The Art and Science of Learning from Data*, 3rd ed. (Agresti and Franklin, 2011)
- *Stats: Modeling the World*, 4th ed. (Bock, Velleman, and De Veaux, 2015)
- *Introduction to Statistics and Data Analysis*, 4th ed. (Peck, Olsen, and Devore, 2012)
- *The Practice of Statistics*, 5th ed. (Starnes et al., 2015).

Advanced Placement program in computer science

In addition to the AP Calculus courses and the AP Statistics course, the College Board also offers two Advanced Placement examinations focused on computer science. At present, the active course is Advanced Placement Computer Science A. This course focuses on object-oriented programming methodology and imperative problem solving, concentrating on problem solving and algorithm development. The course is meant to be the equivalent of a first-semester college-level course in computer science. It also includes the study of the organization of data structures and algorithm design. The course employs a subset of the Java programming language that is described in the course description (College Board 2014b).

Starting in the fall of 2016, the AP program will be introducing a new AP computer science course: AP Computer Science Principles. This course will be focused on introducing students to computer science concepts and ways of exploring data and other forms of information. The focus will be on using the computer to solve problems and address challenging tasks. Students will explore applications of computer science techniques in a variety of disciplines that will enhance their skill and analysis tools for applications in later coursework. These techniques will include working with computer graphics to illustrate a process and manipulating and computing with large data sets in studying trends. This course is structured around big ideas in a manner similar to the new forms of AP Calculus.

Special Schools and Programs for Students in Mathematics

The National Consortium of Secondary STEM Schools (NCSSS), known until 2014 as the National Consortium for Specialized Secondary Schools for Mathematics, Science, and Technology (NCSSSMST), includes more than 100 institutional members with 40,000 students. The goal of the consortium, as its name indicates, is to foster, support, and advance the efforts of specialized schools to attract students and prepare them academically for leadership in the subject areas of mathematics, science, and technology. Some members are boarding schools requiring state residence and highly competitive examinations for entrance, a few are specialized local high schools, and others are regional centers that students may attend for a half or full day for a single year. A look at the Home and Program sections of the organization's website gives an idea of its publications and the breadth of topics discussed at its annual conference (NCSSS 2015).

Programs for K–12 students

In addition to the types of specialized schools described above, students can take advantage of advanced mathematics programs through public and private schools, universities, or other organizations. The first type of program is a diploma-based program that follows an international curriculum managed by the International Baccalaureate Organization (IBO), headquartered in The Hague, The Netherlands. More than 1.3 million students were enrolled in some type of IB program worldwide in 2014. In the United States, 1,675 schools are authorized to offer the IBO Program in some form; more than 5,000 schools are authorized worldwide in more than 140 countries. Of the 1,675 schools in the United States, 869 offer the Diploma Program, a demanding two-year precollege program that leads to examinations and is designed for students who are 16 to 19 years of age. The remaining schools offer the Middle Years Program (589) or the Primary Years Program (487), both of which are designed for younger students. In 2012, the IBO initiated the Career-Related

Program, aimed at meeting the needs of students engaged in career-related programs. This program is currently offered in 72 schools in the United States (IBO 2015).

A second type of special program is a university-centered program offered in two formats as a summer program in mathematics for very capable secondary students. The first format follows a model initiated by the late Julian Stanley at Johns Hopkins University in the 1970s, identifying talent in the upper elementary or middle school grades and offering accelerated courses (usually in the summer but sometimes during the school year) and online courses to enable those students to study more advanced mathematics as well as other disciplines at a younger age. Today, the program serves gifted students in K–grade 12 through a wide range of programs (Johns Hopkins University 2015).

The second format for summer mathematics enhancement follows a model initiated around the same time by Arnold Ross at Notre Dame University. In the Ross Program, students are taught mathematics in a different way from the approach that they would normally be exposed to in school. They are expected to solve problems and deduce propositions in somewhat the same manner as a professional mathematician—by working through the problems on their own or in collaborative groups with some outside hints from mentors. The Ohio State University Ross Program, modeled on Notre Dame's program, recruits both regionally and nationally, and opportunities are available for students across the entire nation (Ross Mathematics Program 2015). Boston College offers a summer program along the same lines for high school students, called the Boston College Math Experience. It covers enrichment topics not in the high school curriculum. In the summer of 2015, the topic was non-Euclidean geometry. Students are also allowed to take coursework alongside the Experience topic. In 2015, several students took linear algebra at the same time that they were completing the Experience program (Boston College Math Experience 2015). Many other universities offer variants of such programs.

A third approach to mathematics enrichment and advancement comes through mathematics clubs. The largest organization of mathematics clubs in the United States is Mu Alpha Theta, founded in 1957. Mu Alpha Theta has more than 2,200 high school and community college chapters and more than 108,000 student members across the United States. Its purpose is to stimulate interest in mathematics by providing recognition of superior mathematical scholarship in students. In addition to holding regional meetings and an annual national meeting, Mu Alpha Theta also publishes a newsletter and provides several other resources for its student members (Mu Alpha Theta 2015).

Programs for undergraduate and graduate students

The National Science Foundation (NSF) funds a large number of research opportunities for undergraduate students through its Research Experiences for Undergraduates (REU) program (NSF 2015). A REU site consists of a group of 10 to 15 undergraduates who work on aspects of the active research programs of the sponsoring college or university. Each student is associated with a specific research project and works closely with the faculty and other researchers involved in that program. Students are granted stipends and, in many instances, assistance with housing and travel. Undergraduate students supported with NSF funds must be citizens or permanent residents of the United States or its territories. In 2014–15, 47 REU sites with research opportunities were available in mathematics. A list of the REU sites for 2015–16 can be found at NSF's REU website (NSF 2015).

The National Institutes of Health offers another federally funded summer institute for undergraduates: the Summer Institute for Training in Biostatistics (SIBS). This program provides a six- to seven-week training course for undergraduates and beginning graduate students interested in learning about biostatistics. In the summer of 2015, SIBS institutes were offered on eight different university campuses nationwide. Additional information on the program can be found at the SIBS website: https://www.nhlbi.nih.gov /research/training/summer-institute-biostatistics-t15.

| **Mathematics Competitions for K–12 Students** | Mathematics competitions in the United States are voluntary for both individuals and schools. Some middle schools and high schools have mathematics teams, often competing in events operated by local professional organizations. Descriptions follow of the larger competitions of national scope: |

- **MATHCOUNTS.** The National Society of Professional Engineers, the CAN Foundation, and NCTM founded MATHCOUNTS in 1982 to increase interest and involvement in mathematics and to assist in developing a technologically literate population. The MATHCOUNTS Foundation now operates the competition. Sponsors include the National Society of Professional Engineers, the National Council of Teachers of Mathematics, CAN, Raytheon Company, Northrup Grumman Foundation, the U.S. Department of Defense, Phillips 66, Texas Instruments Incorporated, 3M Foundation, Art of Problem Solving (AoPS), and Next Thought.

  Participation is restricted to students in grades 6, 7, and 8. MATHCOUNTS activities involve school-based club and competition activities. At a broader level are competitions at the chapter level of the Society of Professional Engineers, the state level, and the national level. Each year, more than 250,000 students in 7,000 schools are exposed to MATHCOUNTS materials and programs, and more than 140,000 students from all 50 states, U.S. territories, and the Department of Defense–related school systems participate in the national competition at some level (MATHCOUNTS 2015).

- **American Mathematics Competitions** (AMC). The AMC contests, centered at the University of Nebraska–Lincoln, involved more than 350,000 participants in 2010. These participants account for 20% of the high schools in the country each year. The AMC competitions began in 1950 under the sponsorship of the Mathematical Association of America (MAA) and the Society of Actuaries as the American High School Mathematics Examination (AHSME) for students in grades 9–12. This program, administered by the MAA and principally funded by the Akamai Foundation with the support of 19 other mathematics organizations, has evolved into a series of examinations spanning the range from middle school or junior high school through grade 12. The original AHSME examination is now called the AMC12. Over time, as other organizations became involved, new competitions were added. In 1985, an exam for students below grade 9, the American Junior High School Mathematics Examination (AJHSME), now called the AMC8, was initiated. In 2000, the AMC10, an exam for students below grade 11, was launched (MAA 2015c). In addition to being a freestanding competition, the AMC12 is the first examination in a series of examinations that leads to the selection of the U.S. competitors for the Mathematical Olympiad. The highest scorers on the AMC12 become eligible to participate in the United States of America Mathematical Olympiad (USAMO), a six-question, six-hour exam that is used to determine the U.S. team members for the International Mathematical Olympiad (IMO). The AMC also operates a summer program for qualifying students (MAA 2015d).

In the fall of 2014, the AMC8 involved 146,425 students from 2,356 schools. In the spring of 2015, a total of 186,400 students completed either the AMC10 or AMC12. From the AMC12, a total of 11,643 students moved forward to participate in the American Invitational Mathematics Examination (AIME) (MAA 2015a). From the AIME results, 294 students were selected to sit for the USAMO examination. From this group, 12 students emerged to form the pool from which the final 6 students were selected for the U.S. team for the International Mathematical Olympiad (MAA 2015b, 2015c).

- **The American Statistical Association/National Council of Teachers of Mathematics Poster Competition and Project Competition**. The ASA/NCTM Joint Committee on the Curriculum in Statistics and Probability sponsors an open competition each year for posters from students in K–grade 12 and statistical projects from students in grades 7–12 (ASA 2015a). Posters are to be developed and constructed on flat poster board by a student or group of students. For K–grade 3, there is no limit on the number of students in the group. Above that level, the maximum number of students who may work on a poster is four. The subject matter is of the student's or students' own choosing, but the submitted posters are assessed on "demonstration of important relationships and patterns, obvious conclusions, and ability to stand alone, even without the explanatory paragraph on the back" of the poster board. The posters are classified into grade intervals for judging: K–3, 4–6, 7–9, and 10–12. Original research studies and results are accepted along with the data, statement of purpose, and method of collection of the data.

  Projects, like the posters, have subject matter that is selected by the participants themselves. A group of students competing together may not have more than four students working on a project. The students may have some adult guidance, but the amount of guidance must be detailed in the project write-up. The statistical methods in the projects are limited to what might be found in an introductory statistics class. Statisticians and teachers use the stages of a statistical experiment in evaluating the projects.

  The entry rules and regulations for both posters and projects can be found at http://www.amstat.org/education/posterprojects/posterrules.cfm. Additional information and education available for teachers and students at the ASA website is detailed at http://amstat.org/education.

- **The Math League**. Founded in 1977, the Math League specializes in mathematics contests, books, and computer software designed to stimulate interest and confidence in mathematics for students from fourth grade through high school. In recent years, more than one million students from the United States and China have participated in Math League contests each year. These contests involve students in individual and team-based competitions. Contest problems are designed to cover a range of mathematical knowledge for each grade level and require no additional knowledge of mathematics beyond the grade level that they test (Math League 2015).

- **The American Regions Mathematics League (ARML).** ARML, begun in 1976 as the Atlantic Region Mathematics League, organizes a competition of teams of high school students who represent their school, local area, state, or country (outside the United States). This contest takes place during November and February of a school year. Teams of students from different schools compete in a contest to solve a set of honors-level problems in a 45-minute period of time. The papers are then mailed in and evaluated by a team of judges. A national competition, which takes place toward the end of the school year, occurs at three sites. In May 2015, more than 2,000 students from 134 teams representing schools or regions participated in the national competition (ARML 2015; Steve Condie, personal communication with the author, September 13, 2015).

- **High School Mathematical Contest in Modeling (HiMCM).** Sponsored by the Consortium for Mathematics and its Applications (COMAP) in conjunction with the MAA, NCTM, the Institute for Operations Research and the Management Sciences (INFORMS), and the Society for Industrial and Applied Mathematics (SIAM), HiMCM is an open competition for individuals or teams of up to four secondary students attending the same school. With this formulation, HiMCM allows or teams of one homeschooled student or any number up to four attending the same school. The competition was designed to provide students with an opportunity to work as members of a team on a mathematical modeling problem testing their capabilities to merge their mathematical knowledge with knowledge of a particular context, their ability to use technology where necessary, and their capability to write a complete and coherent explanation of their model and demonstrate its value in determining one or more solutions to the problem posed. Students have access to all forms of inanimate assistance and are monitored by a teacher from the school. They are allowed 36 consecutive hours from the specified beginning time to select one of two problems provided by the competition and develop a model and carry out their explication of that model's relevance to the situation posed. Sample problems and student responses can be found at the HiMCM website: http://www.comap.com/highschool/contests/himcm/index.html (COMAP 2015b; SIAM 2016).

- **International Mathematical Modeling Challenge (IM²C).** IM²C is a new competition that branches off the HiMCM competition described above. IM²C takes competitions for secondary students in the direction of a modeling Olympiad. Like HiMCM, IM²C involves teams of students working on developing and representing the effectiveness of their models for resolving problems presented to them in a contextualized setting. So far, IM²C is operating in a pilot mode as the contest ramps up to a full-fledged international contest. The first contest took place in April of 2015. Ten countries were selected for the pilot process. Each country was limited to two teams, each with up to four students. Each team was monitored by a faculty member associated with the education institution that the students attended. The students had a five-day period between April 15 and May 15, 2015. At the conclusion of those five days, they had to submit their model and explanation of how it solved the problem presented. In 2015, the problem involved movie scheduling, and students were asked to model various aspects of efficiency associated with a movie studio releasing a movie on time under a variety of constraints.

Sample responses consisting of teams' models and explanations can be viewed at the project's website: http://immchallenge.org. Various parts of the website offer glimpses of the work of the students on the problems presented and the instructions given. In future years, the contest will be opened up to teams from a larger number of countries, and additional rules will be developed to increase participation, perhaps with a national modeling contest in the fall to select the best teams and an opportunity in the spring for each of these teams to consider a second modeling challenge to determine the international competition, or International Modeling Olympiad, portion of the contest (COMAP 2015a).

Mathematics Competitions for Undergraduate Students

A number of competitions are open to undergraduates in the United States. The following are among the most prominent:

- **Student Mathematics League (SML).** The SML competition is for students enrolled in two-year colleges. Originally founded in 1970 by Nassau Community College in New York, this twice-annual competition came under the sponsorship of AMATYC in 1981. The SML involves more than 8,000 two-year college students from 165 colleges in 35 states and Bermuda in its annual cycle of two examinations, one in November and the other in March. In each academic year, a team and an individual champion are determined, and a scholarship is awarded as well. In addition to the national results, regional individual and team standings are determined for the eight regional sectors of the United States in the AMATYC governing structure. Each set of these regional results, individual and team, ranks the top 5 entrants, individual or team. The examinations are based on the standard syllabus in college algebra and trigonometry and may involve precalculus-level algebra, trigonometry, synthetic and analytic geometry, and probability. All questions are short-answer or multiple choice (AMATYC 2015).

- **Mathematical Contest in Modeling (MCM)** and the **Interdisciplinary Contest in Modeling (ICM).** At the undergraduate level, the MCM and ICM contests are organized and sponsored by COMAP, along with the MAA, INFORMS, SIAM, Department of Defense Analysis at the Naval Postgraduate School, Chinese Society for Industrial and Applied Mathematics, and Two Sigma, a quantitative financial analysis firm. These two contests are for teams of students at the undergraduate level and below. Students can compete alone or in teams of two or three members, all of whom are attending the same high school, college, or university. The actual competition is held in late January or early February. Contestants have exactly four days, or 96 consecutive hours, to complete a task requiring them to solve one of six problems presented to them at the beginning of the time period. Three of the problems are mathematical modeling problems and the other three are interdisciplinary modeling problems, each requiring some knowledge of another discipline in addition to mathematics. In 2016, these latter three problems will focus on network science, environmental science, and policy.

Participants' problem solutions are then submitted and judged by a panel of experts who are skilled in modeling and have backgrounds in the areas applicable to the problems featured in the competition. By April of the contest year, the results of the competition are posted on the COMAP website. Selected outstanding papers for each problem are then published in the UMAP Journal. In 2015, 2,137 teams from 504 institutions in seven countries participated in the ICM portion of the contests, and 7,636 teams from 971 institutions participated in the MCM portion (COMAP 2015c; Chris Arney et al. 2015; Kristen Arney et al. 2015).

- **Undergraduate Statistics Project Competition (USPROC; Consortium for the Advancement of Undergraduate Statistics [CAUSE]).** This competition is open to undergraduate students studying statistics at the introductory or intermediate level. Teams of one or two students select a topic requiring the application of statistics to solve and then carry out the appropriate statistical analysis on the data collected and report the conclusions. The contest is held every other year, and cash prizes are awarded to the top three teams. The rules and regulations for the contest, and its two subcategories, can be found at https://www.causeweb .org/usproc/. Additional information about CAUSE can be found at https:// www.causeweb.org (CAUSE 2015).

- **William Lowell Putnam Mathematical Competition.** The Putnam Competition is for undergraduate mathematics students and is administered annually by the MAA on the first Saturday in December. The competition is perhaps the most rigorous and prestigious mathematics examination held each year. This examination can be entered by individuals or by three-person teams representing their college or university. Its difficulty is measured by the fact that every year a third of the participants, all outstanding students, receive marks of zero. The Putnam Competition marked its 75th anniversary in December 2014. In this competition, 4,230 individuals and 431 three-person teams competed from 557 colleges and universities in the United States and Canada. Problems and solutions for the 2014 competition are described by Klosinski, Alexanderson, and Krusemeyer (2015).

Chapter 9: Resources

Teachers, administrators, parents, and the public often want access to information on professional mathematics organizations in the United States that are involved in supporting the improvement of mathematics education. The first section of this chapter identifies membership organizations with both closed and open membership and gives their contact points, a brief overview, and any regular publications that they may produce. The second section provides a listing of publishing companies currently involved in producing textual or electronic materials for students at the pre-K–grade 12 levels in the United States. *References to the organizations in this chapter are given in full and will not be repeated in the Bibliography.*

Professional Organizations in Mathematics Education

The following list of U.S. professional organizations in mathematics education is separated according to whether membership is by appointment (closed) or is open to anyone on the basis of their own desires. Organizations in the list have a total or partial focus on the furtherance of the mathematics education of U.S. students.

Closed-membership organizations

Members of the following two organizations are selected or appointed by the organizations themselves or the National Research Council:

- **Conference Board of the Mathematical Sciences (CBMS)**

  e-mail: rosier@georgetown.edu; website: www.cbmsweb.org

  Founded in 1960, CBMS is an umbrella organization consisting of the major professional societies in the mathematical sciences in the United States and composed of the CBMS Executive Committee and the presidents and executive directors of the member societies. Its purpose is to promote understanding and cooperation among the national professional organizations in mathematics so that they can work together, supporting one another in research, the improvement of education, and the expansion of the mathematical sciences. The following societies belong: American Mathematical Association of Two-Year Colleges (AMATYC), American Mathematical Society (AMS), Association of Mathematics Teacher Educators (AMTE), American Statistical Association (ASA), Association for Symbolic Logic (ASL), Association for Women in Mathematics (AWM), Association of State Supervisors of Mathematics (ASSM), Benjamin Banneker Association (BBA), Institute of Mathematical Statistics (IMS), Institute for Operations Research and the Management Sciences (INFORMS), Mathematical Association of America (MAA), National Association of Mathematicians (NAM), National Council of Supervisors of Mathematics (NCSM), National Council of Teachers of Mathematics (NCTM), Society for Industrial and Applied Mathematics (SIAM), Society of Actuaries (SOA), and TODOS: Mathematics for All (TODOS).

- **United States National Commission on Mathematics Instruction (USNC/MI)**

  website: http://sites.nationalacademies.org/PGA/biso/ICMI/index.htm

  The U.S. National Academy of Sciences (NAS) is the national adhering body to the International Commission on Mathematical Instruction (ICMI). Acting through the NRC, the USNC/MI, which was founded in 1978, conducts the work of the ICMI in the United States and fosters other international collaborations in mathematics education (ICMI 2015). The NRC Board of Mathematical Sciences, CBMS, and NCTM suggest nominees for the USNC/MI.

Open-membership organizations

Numerous mathematics organizations in the United States have open memberships—that is, the members self-select and join on their own, often through the payment of an annual membership fee. The following open-membership organizations have as their primary focus mathematics in kindergarten through grade 12:

- **National Council of Supervisors of Mathematics (NCSM)** (founded 1969)
  e-mail: office@mathedleadership.org; website: www.mathedleadership.org/
  journal: *Journal of Mathematics Education Leadership*

- **National Council of Teachers of Mathematics (NCTM)** (founded 1920)
  e-mail: nctm@nctm.org; website: www.nctm.org
  journals: *Teaching Children Mathematics, Mathematics Teaching in the Middle School, Mathematics Teacher, Mathematics Teacher Educator* (joint publication with AMTE), *Journal for Research in Mathematics Education*

- **School Science and Mathematics Association (SSMA)** (founded 1902)
  e-mail: office@ssma.org; website: www.ssma.org
  journal: *School Science and Mathematics*

- **Women and Mathematics Education (WME)** (founded 1978)
  e-mail: m.melissa.hosten@gmail.com; website: www.wme-usa.org

The following open-membership organizations have as their primary focus mathematics at the postsecondary level:

- **American Mathematical Association of Two-Year Colleges (AMATYC)** (founded 1974)
  e-mail: amatyc@amatyc.org; website: www.amatyc.org
  journal: *MathAMATYC Educator*

- **American Mathematical Society (AMS)** (founded 1888)
  e-mail: ams@ams.org; website: www.ams.org
  journals: *Notices of the American Mathematical Society, Bulletin of the American Mathematical Society*

- **American Statistical Association (ASA)** (founded 1839)
  e-mail: asainfo@amstat.org; website: www.amstat.org
  journal: *The American Statistician, Chance, Significance*, and others devoted to research in statistics

- **Mathematical Association of America (MAA)** (founded 1915)
  e-mail: maahq@maa.org; website: www.maa.org
  journals: *The American Mathematical Monthly, College Mathematics Journal, Mathematics Magazine, Math Horizons*

- **National Association of Mathematicians (NAM)** (founded 1969)
  e-mail: leon.woodson@morgan.edu; website: www.nam-math.org
  journal: *NAM Newsletter*

The following open-membership organizations have as their primary focus special groups in mathematics education:

- **Association of Mathematics Teacher Educators (AMTE)** (founded 1993)
  e-mail: hendrixt@meredith.edu; website: www.amte.net
  journals: *Mathematics Teacher Educator, AMTE Connections*

- **Benjamin Banneker Association (BBA)** (founded 1986)
  e-mail: president@bannekermath.org; website: www.bannekermath.org

- **National Association of Community College Teacher Education Programs (NACCTEP)** (founded 2003)
  e-mail: nacctep@riosalado.edu; website: www.nacctep.org

- **Research Council on Mathematics Learning (RCML)** (founded 1974)
  e-mail: william.speer@unlv.edu; website: web.unlv.edu/RCML/
  journal: *Investigations in Mathematics Learning*

- **Special Interest Group for Research in Mathematics Education (SIGRME)**
  e-mail: mshaugh@umich.edu; website: www.sigrme.org

- **TODOS: Mathematics for ALL** (founded 2003)
  e-mail: requests@todos-math.org; website: www.todos-math.org
  journal: *Noticias de TODOS*

Textbook and Electronic Media Publishers, K–Grade 12

Textbooks and electronic media for K–grade 12 are not listed in *Books in Print*, and currently available textbooks are not likely to be listed even in online bookstore catalogs. For this reason, this list of publishers, with their locations and URLs, is provided to assist those who might be interested in obtaining more information about textbooks and electronic media for K–grade 12 and other curricular materials used in the United States. These publishers and their college publishing counterparts often have auxiliary materials available online. Additional information can be found at the website of the State Instructional Materials Review Association (SIMRA): www.simra.us. SIMRA represents the nation's leading developers of instructional materials, technology-based curricula, and assessments. A committee of SIMRA, the Advisory Commission on Textbook Specifications, provides information concerning book and instructional media at the state level. SIMRA is composed of the individuals responsible for the selection and administration of school textbook and instructional media policies for the different states that have state adoption processes. These states control such a large percentage of the U.S. school and media purchases that their decisions shape, to a great extent, the actual content and coverage sequences found in U.S. schools for K–grade 12.

Textual materials

The following list identifies major publishers of textual materials in United States:

- **Amsco School Publications,** a division of Perfection Learning, Logan, IA 51546; www.amscopub.com

- **Carnegie Learning**, Pittsburgh, PA 15219; www.carnegielearning.com
- **CORD Communications**, Waco, TX 76702; www.cord.org
- **CPM Educational Program**, Elk Grove, CA 95758; www.cpm.org
- **Curriculum Research and Development Group**, Honolulu, HI 96822; www.hawaii.edu/crdg
- **Macmillan/McGraw-Hill**, Columbus, OH 53272; www.mheonline.com
- **Harcourt/Holt/McDougal/Houghton Mifflin**, Boston, MA 02116; www.harcourtschool.com
- **It's About Time, Inc.,** Armonk, NY 10504; www.its-about-time.com
- **Kendall/Hunt Publishing Company**, Dubuque, IA 52004; www.kendallhunt.com
- **Key Curriculum Press**, Emeryville, CA 94608; www.keypress.com
- **Pearson/Addison Wesley/Dale Seymour/Prentice Hall/Scott-Foresman**, Boston, MA 02116; www.pearsonschool.com
- **Sadlier-Oxford,** New York, NY 10005; www.sadlier-oxford.com

Non-text digital supplementary materials

The following list sorts and presents by grade band the major sources of non-text digital supplementary materials in the United States:

K–grade 2

- **IXL Learning**, San Mateo, CA 94404; www.IXL.com
- **Coolmath Games**, New York, NY 10168; www.coolmath.com
- **Math Playground LLC,** www.mathplayground.com/

Grades 3–5

- **IXL Learning**, San Mateo, CA 94404; www.IXL.com
- **Coolmath Games**, New York, NY 10168; www.coolmath.com

Grades 6–8

- **Kahn Academy**, Mountain View, CA 94097; www.khanacademy.org
- **StudyIsland**, Suite 300, 200 Tower, Bloomington, MN 55437; studyisland.com

Grades 9–12

- **Kahn Academy,** Mountain View, CA 94097; www.khanacademy.org
- **PurpleMath**, Palatine IL 60078; PurpleMath.com

Bibliography

Academic Benchmarks. *Common Core State Standards Adoption Map.* Cincinnati: Academic Benchmarks, 2015. Accessed December 12, 2015. http://academicbenchmarks.com/common -core-state-adoption-map/

ACT. *The Condition of College and Career Readiness 2015.* Iowa City, Iowa: ACT, 2015. Accessed September 7, 2015. http://www.act.org/research/policymakers/cccr15/pdf/CCCR15 -NationalReadinessRpt.pdf

Agresti, Alan, and Christine A. Franklin. *Statistics: The Art and Science of Learning from Data.* 3rd ed. Boston: Pearson Education, 2011.

Aliaga, Martha, George Cobb, Carolyn Cuff, Joan Garfield, Rob Gould, Robin Lock, Tom Moore, Allan Rossman, Bob Stephenson, Jessica Utts, Paul Velleman, and Jeff Witmer. *Guidelines for Assessment and Instruction in Statistics Education: College Report.* Alexandria, Va.: American Statistical Association, 2007. Accessed August 17, 2015. http://www.amstat.org/education/gaise/

American Mathematical Association of Two-Year Colleges (AMATYC). *Crossroads in Mathematics: Standards for Introductory College Mathematics.* Memphis, Tenn.: AMATYC, State Technical Institute at Memphis, 1995. Accessed August 17, 2015. http://www.amatyc.org /Crossroads/CROSSROADS/V1/index.htm

———. *Beyond Crossroads: Implementing Mathematics Standards in the First Two Years of College.* Memphis, Tenn.: AMATYC, Southwest Tennessee Community College, 2006. Accessed August 17, 2015. http://beyondcrossroads.amatyc.org/doc/PDFs/BCAll.pdf

———. *Student Mathematics League* (SML). Memphis, Tenn.: AMATYC, Southwest Tennessee Community College, 2015. Accessed December 19, 2015. http://www.amatyc .org/?StudentMathLeague

American Regions Math League (ARML). *The Official American Regions Mathematics League Web Page.* Accessed December 19, 2015. http://www.arml.com/arml_2015/page/index .php?page_type=public&page=home

American Statistical Association (ASA). *Curriculum Guidelines for Undergraduate Programs in Statistical Sciences.* Alexandria, Va.: ASA, 2014. Accessed December 8, 2015. http://www .amstat.org/education/pdfs/guidelines2014-11-15.pdf

———. *The American Statistical Association/National Council of Teachers of Mathematics Poster Competition and Project Competition.* Alexandria, Va.: ASA, 2015a. Accessed December 22, 2015. http://www.amstat.org/education/posterprojects/posterrules.cfm

Arney, Chris, Kathryn Coronges, Tina Hartley, Rod Sturdivant, and Robert Ulman. "Judges' Commentary: Managing Human Capital in Organizations." *UMAP Journal* 36 (Summer 2015): 137–49.

Arney, Kristen, and Jessica Libertini. "Judges' Commentary: Managing Human Capital in Organizations." *UMAP Journal* 36 (Summer 2015): 171–88.

ARTIST Project. *Welcome to the ARTIST Web Site.* Minneapolis: University of Minnesota, 2015. Accessed December 22, 2015. https://apps3.cehd.umn.edu/artist/index.html

Association of American Publishers. *Educational Standards.* Washington, D.C.: Association of American Publishers, 2015. Accessed December 8, 2015. http://publishers.org/priorities -positions/educational-standards

Aud, Susan, William Hussar, Grace Kena, Kevin Bianco, Lauren Frohlich, Jana Kemp, and Kim Tahan. *The Condition of Education 2011.* Washington, D.C.: National Center for Education Statistics, 2011. Accessed October 16, 2015. https://nces.ed.gov/pubs2011/2011033.pdf

Banilower, Eric R., P. Sean Smith, Iris R. Weiss, Kristen A. Malzahn, Kiira M. Campbell, and Aaron M. Weiss. *Report of the 2012 National Survey of Science and Mathematics Education*. Chapel Hill, N.C.: Horizon Research, 2013.

Barton, Paul E., and Richard J. Coley. *The Family: America's Smallest School*. Princeton, N.J.: Policy Evaluation and Research Center, Educational Testing Service, 2007.

Beaton, Albert E., Ina V. S. Mullis, Michael O. Martin, Eugenio J. Gonzalez, Dana L. Kelly, and Teresa A. Smith. *Mathematics Achievement in the Middle School Years: IEA's Third International Mathematics and Science Study (TIMSS)*. Chestnut Hill, Mass.: TIMSS International Study Center. Boston College, 1997. Accessed August 17, 2015. http://timss.bc.edu

Berliner, David C., and Gene V. Glass. *50 Myths and Lies That Threaten America's Public Schools*: *The Real Crisis in Education*. New York: Teachers College Press, Columbia University, 2014.

Blair, Richelle M., Ellen E. Kirkman, and James W. Maxwell. *Statistical Abstract of Undergraduate Programs in the Mathematical Sciences in the United States: Fall 2010 CBMS Survey*. Providence, R.I.: American Mathematical Society, 2013. Accessed September 2, 2015. http://www.ams.org/profession/data/cbms-survey/cbms2010-Report.pdf

Blank, Rolf, Doreen Langesen, and Adam Petermann. *State Indicators of Science and Mathematics Education: 2007*. Washington, D.C.: Council of Chief State School Officers, 2007.

Bock, David E., Paul F. Velleman, and Richard D. De Veaux. *Stats: Modeling the World*. 4th ed. Boston: Pearson Education, 2015.

Boston College Math Experience. *Boston College Math Experience*. Boston College, Chestnut Hill, Mass.: Boston College Summer Session—BCE, 2015. Accessed December 19, 2015. http://www.bc.edu/schools/summer/bce/academics/bcemath.html

Braswell James S., Anthony D. Lutkus, Wendy S. Grigg, Shari L. Santapau, Brenda Tay-Lim, and Matthew Johnson. *The Nation's Report Card: Mathematics 2000*, Washington, D.C.: U.S. Department of Education, National Center for Education Statistics, 2001. Accessed December 9, 2015. http://nces.ed.gov/nationsreportcard/pdf/main2000/2001517a.pdf

Braun, Henry, Frank Jenkins, and Wendy Grigg. *A Closer Look at Charter Schools Using Hierarchical Linear Modeling*. Washington, D.C.: U.S. Department of Education, National Center for Education Statistics, 2006. Accessed December 17, 2015. http://nces.ed.gov/nationsreportcard/pdf/studies/2006460.pdf

Burstein, Leigh, ed. *The IEA Study of Mathematics III: Student Growth and Classroom Processes*. Oxford, UK: Pergamon Press, 1993.

Bush, George W. *No Child Left Behind. Washington*, D.C.: U.S. Department of Education, Office of the Secretary, 2001. Accessed September 2, 2015. http://eric.ed.gov/?q=George+W.+Bush&ft=on&pg=6&id=ED447608

Bush, Melodye. *ECS State Notes: Compulsory School Age Requirements*. Denver: Education Commission of the States, 2010. Accessed July 29, 2015. http://www.ncsl.org/documents/educ/ECSCompulsoryAge.pdf

California Department of Education. *California 2011 State Report to National Association of State Textbook Administrators*, 2011. Accessed January 1, 2016. http://nasta.org/CA_NASTAStateReport2011.pdf

Campbell, Jay R., Catherine M. Hombo, and John Mazzeo. *NAEP 1999 Trends in Academic Progress: Three Decades of Student Performance*. Washington, D.C.: U.S. Department of Education, National Center for Education Statistics, 2000. Accessed December 9, 2015. https://nces.ed.gov/nationsreportcard/pdf/main1999/2000469.pdf

Campbell, Jay R., Kristin E. Voelkl, and Patricia L. Donahue. *NAEP 1996 Trends in Academic Progress: Achievement of U.S. Students in Science, 1969 to 1996, Mathematics, 1973 to 1996,*

Reading, 1971 to 1996, Writing, 1984 to 1996. Washington, D.C.: National Center for Education Statistics, 1997. Accessed December 9, 2015. http://nces.ed.gov/nationsreportcard//pdf /main1996/97985r.pdf

CAUSE (Consortium for the Advancement of Undergraduate Statistics Education). *USPROC: Undergraduate Statistics Project Competition*. University Park, Pa.: Department of Statistics, Penn State University, 2015. Accessed December 19, 2015. https:// www.causeweb.org

Center for Research on Education Outcomes. *Multiple Choice: Charter School Performance in 16 States*. Stanford, Calif.: CREDO at Stanford University, 2009. Accessed December 17, 2015. http://credo.stanford.edu/reports/MULTIPLE_CHOICE_CREDO.pdf

————. *National Charter School Study: 2013*. Stanford, Calif.: CREDO at Stanford University, 2013. Accessed December 17, 2015. http://credo.stanford.edu/documents/NCSS %202013%20Final%20Draft.pdf

Chapman, Chris, Jennifer Laird, and Angelina Kewal-Ramani. *Trends in High School Dropout and Completion Rates in the United States: 1972–2008*. Washington, D.C.: National Center for Education Statistics, 2010. Accessed July 28, 2015. http://www.nces.ed.gov

Cobb, George. "Teaching Statistics." In *Heeding the Call for Change: Suggestions for Curricular Action*, edited by Lynn Arthur Steen, pp. 3–43. Washington, D.C.: Mathematical Association of America, 1992.

College Board. *College Board Standards for College Success: Mathematics and Statistics*. New York: College Board, 2006. Accessed August 17, 2015. http://www.collegeboard.com/prod _downloads/about/association/academic/mathematics-statistics_cbscs.pdf

————. *Statistics: Course Description*. New York: College Board, 2010. Accessed September 7, 2015. https://secure-media.collegeboard.org/ap-student/course/ap-statistics-2010-course -exam-description.pdf

————. *FAQs about the SAT*. New York: College Board, 2011. Accessed September 7, 2015. https://sat.collegeboard.org/about-tests/sat/faq

————. *AP Calculus AB and AP Calculus BC, Including the Curriculum Framework* (*Effective 2016–2017*). New York: College Board, 2014a. Accessed December 18, 2015. https://secure -media.collegeboard.org/digitalServices/pdf/ap/ap-calculus-curriculum-framework.pdf

————. *Computer Science A: Course Description*. New York: College Board, 2014b. Accessed December 18, 2015. https://secure-media.collegeboard.org/digitalServices/pdf/ap/ap -computer-science-a-course-description.pdf

————. *About AP Scores*. New York: College Board, 2015a. Accessed December 18, 2015. https://apscore.collegeboard.org/scores/about-ap-scores

————. *AP Program Summary Report 2015*. New York: College Board, 2015b.

————. *AP Student Score Distributions, May 2015*. New York: College Board, 2015c.

————. *AP Credit-Granting Recommendations*. New York: College Board, 2015d. Accessed December 18, 2015. https://aphighered.collegeboard.org/setting-credit-placement-policy/ credit-granting-recommendations

————. *SAT*. New York: College Board, 2015e. Accessed September 7, 2015. https://collegereadi-ness.collegeboard.org/sat?navid=tnt-satn

————. *SAT Total Group Profile Report: 2015 College-Bound Seniors*. New York: College Board, 2015f. Accessed September 7, 2015. https://secure-media.collegeboard.org/digitalServices/ pdf/sat/total-group-2015.pdf

————. *Student Score Distributions*. New York: College Board, 2015g.

————. *Trends in College Pricing: 2014.* New York: College Board, 2015h.

Conference Board of Mathematical Sciences (CBMS). *The Mathematical Sciences Curriculum K–12: What Is Still Fundamental and What Is Not.* Washington, D.C.: CBMS, 1982. Accessed August 17, 2015. http://www.mathcurriculumcenter.org/PDFS/CCM/originals/what_is_fundamental_report.pdf

————. *New Goals for Mathematical Science Education.* Washington, D.C.: CBMS, 1984.

————. *The Mathematical Education of Teachers I.* Providence R.I.: American Mathematical Society, 2001. Accessed August 17, 2015. http://www.cbmsweb.org/MET_Document/index.htm

————. *The Mathematical Education of Teachers II.* Providence R.I.: American Mathematical Society, 2012. Accessed August 17, 2015. http://www.cbmsweb.org/MET2/index.htm

Conley, David. T. *The Common Core State Standards: Insight into Their Development and Purpose.* Washington, D.C.: Council of Chief State School Officers, 2014. Accessed October 22, 2015. http://www.ccsso.org/Documents/2014/CCSS_Insight_Into_Development_2014.pdf

Consortium for Mathematics and Its Applications (COMAP). *The International Mathematical Modeling Challenge, IM²C.* Bedford, Mass.: COMAP, 2015a. Accessed December 19, 2015. http://immchallenge.org

————. *High School Mathematical Contest in Modeling (HiMCM).* Bedford, Mass.: COMAP, 2015b. Accessed December 19, 2015. http://www.comap.com/highschool/contests/himcm/about.html

————. *MCM: The Mathematical Contest in Modeling/The ICM: The Interdisciplinary Contest in Modeling.* Bedford, Mass.: COMAP, 2015c. Accessed December 19, 2015. https://www.comap.com/undergraduate/contests/mcm/

Council for the Accreditation of Educator Preparation (previously NCATE). *NCTM 2012 CAEP Standards.* Reston, Va.: National Council of Teachers of Mathematics, 2012. Accessed August 17, 2015. https://www.nctm.org/ncate/

Darling-Hammond, Linda, and Maria E. Hyler. "The Role of Performance Assessment in Developing Teaching as a Profession." *Rethinking Schools* 27 (Summer 2013): 10–15. Accessed December 15, 2015. http://www.rethinkingschools.org/archive/27_04/27_04_darling-hammond_hyler.shtml

Daro, Phil, Gerunda B. Hughes, and Fran Stancavage. *Study of the Alignment of the 2015 NAEP Mathematics Items at Grades 4 and 8 to the Common Core State Standards (CCSS) for Mathematics.* Washington, D.C.: American Institutes for Research, NAEP Validity Studies (NVS) Panel, 2015. Accessed December 11, 2015. http://www.air.org/sites/default/files/downloads/report/Study-of-Alignment-NAEP-Mathematics-Items-common-core-Nov-2015.pdf

Dossey, John A., and Mary M. Lindquist. "The Impact of TIMSS on the Mathematics Standards Movement in the United States." In *Secondary Analysis of the TIMSS Data*, edited by David F. Robitaille and Albert E. Beaton, pp. 63–79. Dordrecht, The Netherlands: Kluwer Academic Publishers, 2002.

Dossey, John A., and Margaret Wu. "International Assessments of Mathematics and Policy." In *Third Handbook of International Mathematics Education*, edited by M. A. (Ken) Clements, Alan Bishop, Christine Keitel, Jeremy Kilpatrick, and Fred Leung, pp. 1009–42. Dordrecht, The Netherlands: Springer, 2013.

Duncan, Arne. *ESEA Blueprint for Reform.* Washington, D.C.: U.S. Department of Education, Office of Planning, Evaluation, and Policy Development, 2010.

Dweck, Carol. S. "Mind-Sets and Equitable Education." *Principal Leadership* 10 (January 2010): 26–29.

edTPA. *About edTPA*. Amherst, Mass.: Pearson, 2015. Accessed November 22, 2015. http://www.edtpa.com/PageView.aspx?f=GEN_AboutEdTPA.html

Every Student Succeeds Act (ESSA). Public Law Number No. 114-95. 114th Congress, 2015.

Fields, R. *Towards the National Assessment of Educational Progress (NAEP) as an Indicator of Academic Preparedness for College and Job Training*. Washington, D.C.: National Assessment Governing Board, 2014.

Finney, Ross, Frank Demana, Bert Waits, Dan Kennedy, and David Bressoud, *Calculus: Graphical, Numerical, Algebraic*. 5th ed. Boston: Pearson/Prentice Hall, 2012.

Franklin, Christine, Anna Bargagliotti, Catherine Case, Gary Kader, Richard Scheaffer, and Denise Spangler. *The Statistical Education of Teachers*. Alexandria, Va.: American Statistical Association, 2015. Accessed August 17, 2015. http://www.amstat.org/education/SET/SET.pdf

Franklin, Christine, Gary Kader, Denise Mewborn, Jerry Moreno, Roxy Peck, Mike Perry, and Richard Scheaffer. *Guidelines for Assessment and Instruction in Statistics Education (GAISE Report): A Pre-K–12 Curriculum Framework*. Alexandria, Va.: American Statistical Association, 2007. Accessed August 17, 2015. http://www.amstat.org/education/gaise/GAISEPreK-12_Full.pdf

Fulkerson, William O. *2012 National Survey of Science and Mathematics Education: Status of Middle School Mathematics*. Chapel Hill, N.C.: Horizon Research, 2013.

Glander, Mark. *Selected Statistics from the Public Elementary and Secondary Education Universe: School Year 2013–14* (NCES 2015-151). Washington, D.C.: National Center for Education Statistics, 2015. Accessed October 16, 2015. http://nces.ed.gov/pubsearch

Glinder, Scott A., Janice E. Kelly-Reid, Farrah B. Mann, and RTI International. *Postsecondary Institutions and Cost of Attendance in 2014–15; Degrees and Other Awards Conferred, 2013–14; and 12-Month Enrollment, 2013–14*. Washington, D.C.: National Center for Education Statistics, 2015. Accessed December 4, 2015. http://nces.ed.gov/pubs2015/2015097rev.pdf

Grigg, Wendy S., Patricia L. Donahue, and Gloria Dion. *The Nation's Report Card: 12th-Grade Reading and Mathematics 2005*. Washington, D.C.: U.S. Department of Education, National Center for Education Statistics, 2007. Accessed December 9, 2015. https://nces.ed.gov/nationsreportcard/pubs/main2007/2007496.asp

Grouws, Douglas A., and Kristin J. Cebulla. "Improving Student Achievement in Mathematics, Part I: Research Findings." *Eric Digest*. Columbus, Ohio: ERIC Clearinghouse for Science, Mathematics, and Environmental Education, December, 2000. Accessed November 20, 2015. http://www.gpo.gov/fdsys/pkg/ERIC-ED463952/pdf/ERIC-ED463952.pdf

Handwerk, Philip, Namrata Tognatta, Richard J. Coley, and Drew H. Gitomer. *Access to Success: Patterns of Advanced Placement Participation in U.S. High Schools*. Princeton, N.J.: Educational Testing Service, 2008.

Hechinger, Fred W. "The U.S. Gets Low Marks in Math." *New York Times*. March 12, 1967, sec. 4, p. 11.

Heck, Daniel J., I. R. Weiss, and Joan D. Pasley. *A Priority Research Agenda for Understanding the Influence of the Common Core State Standards for Mathematics*. Chapel Hill, N.C.: Horizon Research, 2011.

Hiebert, James, Ronald Gallimore, Helen Garnier, Karen Bogard Givvin, Hilary Hollingsworth, Jennifer Jacobs, Angel Miu-Ying Chui, Diana Wearne, Margaret Smith, Nicole Kersting, Alfred Manaster, Ellen Tseng, Wallace Etterbeek, Carl Manaster, Patrick Gonzales, and James Stigler. *Teaching Mathematics in Seven Countries: Results from the TIMSS 1999 Video Study* (NCES 2003–013 Revised). Washington, D.C.: U.S. Department of Education, National Center for Education Statistics, 2003. Accessed November 22, 2015. http://nces.ed.gov/pubs2003/2003013.pdf

Higher Educational Research Institute. *The American Freshman: Expanded National Norms Fall.* Los Angeles: Cooperative Institutional Research Program, Graduate School of Education & Information Studies, University of California, 2010, 2011, 2012, 2013, and 2014. Accessed September 2, 2015. http://www.heri.ucla.edu/tfsPublications.php

Hill, Jason, Christina Stearns, and Chelsea Owens. *Education and Certification Qualifications of Departmentalized Public High School-Level Teachers of Selected Subjects: Evidence from the 2011–12 Schools and Staffing Survey.* Washington, D.C.: U.S. Department of Education, National Center for Education Statistics, 2015.

Hirsch, Christian R., ed. *Perspectives on the Design and Development of School Mathematics Curricula.* Reston, Va.: National Council of Teachers of Mathematics, 2007.

Home School Legal Defense Association (HSLDA). *Homeschooling under Your State Law—Illinois.* Purcellville, Va.: Home School Legal Defense Association, 2015. Accessed December 17, 2015. http://www.hslda.org/hs101/IL.aspx

Hughes, Gerunda B., Phil Daro, Deborah Holtzman, and Kyndra Middleton. *A Study of the Alignment between the NAEP Mathematics Framework and the Common Core State Standards for Mathematics (CCSS-M).* Washington, D.C.: NAEP Validity Studies (NVS) Panel, American Institutes for Research, 2013. Accessed December 11, 2015. http://www.air.org/sites/default/files/downloads/report/NVS_combined__study_1_NAEP_alignment_with_CCSS_0.pdf

Hughes-Hallett, Deborah, William G. McCallum, Andrew M. Gleason, Daniel E. Flath, Patti Frazer Lock, Sheldon P. Gordon, David O. Lomen, David Lovelock, Brad G. Osgood, Andrew Pasquale, Douglas Quinney, Jeff Tecosky-Feldman, Joseph Thrash, Karen R. Rhea, and Thomas W. Tucker. *Calculus: Single Variable.* 6th ed. New York: John Wiley and Sons, 2012.

Husén, Torsten, ed. *International Study of Achievement in Mathematics: A Comparison of Twelve Countries.* Vols. 1–2. Stockholm: Almqvist & Wiksell, 1967.

Hussar, William J., and Tabitha M. Bailey. *Projections of Education Statistics to 2021.* Washington, D.C.: National Center for Education Statistics, 2011. Accessed October 16, 2015. http://nces.ed.gov/pubs2013/2013008.pdf.http://www.nces.ed.gov

———. *Projections of Education Statistics to 2019.* Washington, D.C.: U.S. Department of Education, National Center for Education Statistics, 2013. Accessed September 3, 2015. http://nces.ed.gov/pubs2011/2011017.pdf

———. *Projections of Education Statistics to 2022.* Washington, D.C.: National Center for Education Statistics, 2014. Accessed August 3, 2015. http://nces.ed.gov.

International Baccalaureate Organization (IB). *About the IB.* Bethesda, Md.: International Baccalaureate Organization IB Global Centre. 2015. Accessed December 19, 2015. http://www.ibo.org

International Commission on Mathematics Instruction (ICMI). *A Historical Sketch of ICMI.* Berlin: Secretariat of the International Mathematical Union, 2015. Accessed November 21, 2015. http://www.mathunion.org/icmi/icmi/a-historical-sketch-of-icmi/

Jeub, Chris. "Why Parents Choose Home Schooling." *Educational Leadership* 52 (September 1994): 50–52.

Johns Hopkins University Center for Talented Youth. *Who We Are*. Baltimore: Johns Hopkins Center for Talented Youth, 2015. Accessed December 19, 2015. http://cty.jhu.edu/about/

Kane, Thomas, *Did the Common Core Assessments Cause the Decline in NAEP Scores?* Brookings Institute, Evidence Speaks Series 13, November 5, 2015. Accessed December 10, 2015. http://www.brookings.edu/research/papers/2015/11/05-common-core-assessments-decline -in-naep-scores-kane

Kena, Grace, Lauren Musu-Gillette, Jennifer Robinson, Xiaolei Wang, Amy Rathbun, Sidney Wilkinson-Flicker, Amy Barmer, and Erin Velez. *The Condition of Education 2015*. Washington, D.C.: National Center for Education Statistics, 2015. Accessed August 1, 2015. http://nces.ed.gov

Kifer, Skip. "Opportunities, Talents, and Participation." In *The IEA Study of Mathematics III: Student Growth and Classroom Processes*, edited by Leigh Burstein, pp. 297–307. Oxford, UK: Pergamon Press, 1993.

Kilpatrick, Jeremy, Jane Swafford, and Bradford Findell, eds. *Adding It Up: Helping Children Learn Mathematics*. Washington, D.C.: National Academy Press, 2001.

Klosinski, Leonard F., Gerald L. Alexanderson, and Mark Krusemeyer. "The Seventy-Fifth William Lowell Putnam Mathematical Competition." *American Mathematical Monthly* 122 (October 2015): 715–25.

Kobrin, Jennifer L., and Anne E. Schmidt. *The Research Behind the New SAT*. New York: College Board, 2007.

Koestler, Courtney, Mathew D. Felton, Kristen N. Bieda, and Samuel Otten. *Connecting the NCTM Process Standards and the CCSSM Practices*. Reston, Va: NCTM, 2013.

Larson, Ron, and Bruce H. Edwards. *Calculus: Early Transcendental Functions*. 6th ed. Independence, Ky.: Cengage Learning, 2015.

Latham, Andrew. "Home Schooling." *Educational Leadership* 55(May 1998): 85–86.

Lee, Jihyun, Wendy Grigg, and Gloria Dion, *The Nation's Report Card: Mathematics 2007*. Washington, D.C.: U.S. Department of Education, National Center for Education Statistics, 2007. Accessed December 9, 2015. http://nces.ed.gov/nationsreportcard /pdf/main2007/2007494.pdf

LOCUS. *Levels of Conceptual Understanding in Statistics*. Gainesville, Fla.: LOCUS, 2015. Accessed December 15, 2015. https://locus.statisticseducation.org

Lubienski, Christopher. "Whither the Common Good? A Critique of Home Schooling," *Peabody Journal of Education* 75, nos.1–2 (2000): 207–32.

Lubienski, Christopher A., and Sarah Theule Lubienski. *The Public School Advantage: Why Public Schools Outperform Private Schools*. Chicago: University of Chicago Press, 2013.

Lutzer, David J., Stephen B. Rodi, Ellen E. Kirkman, and James W. Maxwell. *Statistical Abstract of Undergraduate Programs in the Mathematical Sciences in the United States: Fall 2005 CBMS Survey*. Providence, R.I.: American Mathematical Society, 2007. Accessed September 2, 2015. http://www.ams.org/cbms/CBMS2005-Report.pdf

Madeloni, Barbara, and Rachel Hoogstraten. "The Other Side of Fear." *Schools: Studies in Education* 10 (Spring 2013): 7–19.

Malzahn, Kristen A. *2012 National Survey of Science and Mathematics Education: Status of Elementary School Mathematics*. Chapel Hill, N.C.: Horizon Research, 2013.

Math League. *About the Math League*. Oviedo, Fla.: Math League, 2015. Accessed December 19, 2015. http://mathleague.org

MATHCOUNTS. *MATHCOUNTS Competition Series*. Alexandria, Va.: MATHCOUNTS Foundation, 2015. Accessed December 19, 2015. https://www.mathcounts.org/programs /competition-series

Mathematical Association of America (MAA). *Committee on the Undergraduate Program in Mathematics (CUPM): Undergraduate Programs and Courses in the Mathematical Sciences*. Washington, D.C.: Mathematical Association of America, 2004a. Accessed August 17, 2015. http://www.maa.org/cupm/about.html

———. *Discussion Papers about Mathematics and the Mathematical Sciences in 2010: What Should Students Know?* Washington, D.C.: MAA, Committee on the Undergraduate Program in Mathematics (CUPM), 2004b. Accessed August 17, 2015. http://www.maa.org /cupm/about.html

———. *INGenIOuS Project*. Washington, D.C.: Project INGenIOuS, MAA, 2014. Accessed August 17, 2015. http://www.maa.org/sites/default/files/pdf/ingenious/INGenIOuS -report.pdf

———. *CUPM Curriculum Guide to Majors in the Mathematical Sciences: 2015*. Washington, D.C.: MAA, Committee on the Undergraduate Program in Mathematics (CUPM), 2015a. Accessed August 17, 2015. http://www.maa.org/sites/default/files/pdf/CUPM/pdf /CUPMguide_print.pdf

———. *American Invitational Mathematics Examination (AIME)*. Mathematical Association of America, 2015b. Accessed December 19, 2015. http://www.maa.org/math-competitions /amc-contests/invitational-competitions

———. *American Mathematics Competitions (AMC)*. Washington, D.C.: Mathematical Association of America, 2015c. Accessed December 19, 2015. http://www.maa.org/math -competitions

———. *American Mathematics Competitions (USAMO/USAJMO)*. Mathematical Association of America, 2015d. Accessed December 19, 2015. http://www.maa.org/math-competitions /amc-contests/invitational-competitions

Mathematical Sciences Education Board. *Everybody Counts*. Washington, D.C.: National Academy Press, 1989. Accessed August 17, 2015. http://www.nap.edu/catalog/1199/everybody -counts-a-report-to-the-nation-on-the-future

———. *Reshaping School Mathematics*. Washington, D.C.: National Academy Press, 1990. Accessed August 17, 2015. http://www.nap.edu/catalog/1498/reshaping-school-mathematics-a-philosophy-and-framework-for-curriculum

McKnight, Curtis C., F. Joe Crosswhite, John A. Dossey, Edward Kifer, Jane O. Swafford, Kenneth J. Travers, and Thomas J. Cooney. *The Underachieving Curriculum: Assessing U.S. School Mathematics from an International Perspective*. Champaign, Ill.: Stipes Publishing, 1987. Accessed August 17, 2015. http://files.eric.ed.gov./fulltext/ED297930.pdf

McLeod, Douglas B., Robert E. Stake, Bonnie P. Schappelle, Melissa Mellissinos, and Mark J. Gierl. "Setting the Standards: NCTM's Role in the Reform of Mathematics Education." In *Bold Ventures: Case Studies of U.S. Innovations in Mathematics Education*, vol. 3, edited by Senta A. Raizen and Edward D. Britton, pp. 13–132. Dordrecht, The Netherlands: Kluwer Academic, 1996.

Mehta, Jal, and Doctor, Joe. "Raising the Bar for Teaching." *Phi Delta Kappan* 94 (April 2013): 8–13.

Mikulecky, Marga. *School Attendance Age Limits*. Denver, Colo.: Education Commission of the States, 2013. Accessed July 29, 2015. http://www.nces.ed.gov

Moore, Thomas J. *Teaching Statistics: Resources for Undergrad Instructors.* Washington, D.C.: American Statistical Association; Mathematical Association of America, 2001.

Mu Alpha Theta. *The Purpose of Mu Alpha Theta.* Norman, Okla.: Mu Alpha Theta, University of Oklahoma, 2015. Accessed December 19, 2015. http://www.mualphatheta.org/index .php?about-us

Mullis, Ina V. S., John A. Dossey, Mary A. Foertsch, Lee R. Jones, and Claudia A. Gentile. *Trends in Academic Progress: Achievement of U.S. Students in Science, 1969–70 to 1990, Mathematics, 1973 to 1990, Reading, 1971 to 1990, Writing, 1984 to 1990.* Washington, D.C.: U.S. Department of Education, National Center for Education Statistics, 1991a. Accessed December 10, 2015. http://files.eric.ed.gov/fulltext/ED338720.pdf

Mullis, Ina V. S., John A. Dossey, Eugene H. Owen, and Gary W. Phillips. *The State of Mathematics Achievement: NAEP's 1990 Assessment of the Nation and the Trial State Assessment.* Princeton, NJ: Educational Testing Service and Washington, D.C.: U.S. Department of Education, National Center for Education Statistics, 1991b.

Mullis, Ina V. S., Michael O. Martin, Albert E. Beaton, Eugenio J. Gonzalez, Dana L. Kelly, and Teresa A. Smith. *Mathematics Achievement in the Primary School Years: IEA's Third International Mathematics and Science Report.* TIMSS International Study Center. Boston College, Chestnut Hill, Mass., 1997. Accessed August 17, 2015. http://timss.bc.edu

———. *Mathematics and Science Achievement in the Final Year of Secondary School: IEA's Third International Mathematics and Science Study.* Chestnut Hill, Mass.: Boston College, 1998. Accessed November 20, 2015. http://timss.bc.edu/timss1995i/TIMSSPDF/C _full.pdf

Mullis, Ina V. S., Michael O. Martin, Pierre Foy, and Alka Arora. *TIMSS 2011 International Results in Mathematics.* Chestnut Hill, MA: TIMSS & PIRLS International Study Center, Boston College, 2012. Accessed November 22, 2015. http://timssandpirls.bc.edu/timss2011 /downloads/T11_IR_Mathematics_FullBook.pdf

Mullis, Ina V. S. Michael O. Martin, Pierre Foy, John F. Olson, Corinna Preuschoff, Ebru Erberber, Alka Arora, and Joseph Galia. *TIMSS 2007 International Mathematics Report: Findings from IEA's Trends in International Mathematics and Science Study at the Fourth and Eighth Grades.* Chestnut Hill, Mass.: TIMSS & PIRLS International Study Center, Boston College, 2008a. Accessed November 20, 2015. http://timss.bc.edu/TIMSS2007/PDF/TIMSS2007 _InternationalMathematicsReport.pdf

Mullis, Ina V. S., Michael O. Martin, Eugenio J. Gonzalez, and Steven J. Chrostowski. *TIMSS 2003 International Mathematics Report: Findings from IEA's Trends in International Mathematics and Science Study at the Fourth and Eighth Grades.* Chestnut Hill, Mass.: TIMSS & PIRLS International Study Center, Boston College, 2004. Accessed November 20, 2015. http://timss.bc.edu/PDF/t03_download/T03INTLMATRPT.pdf

Mullis, Ina V. S., Michael O. Martin, Eugenio J. Gonzalez, Kelvin D. Gregory, Robert A. Garden, Kathleen M. O'Connor, Steven J. Chrostowski, and Teresa A. Smith. *TIMSS 1999 International Mathematics Report: Findings from IEA's Repeat of the Third International Mathematics and Science Study at the Eighth Grade.* Chestnut Hill, Mass.: TIMSS & PIRLS International Study Center, Boston College, 2000. Accessed August 17, 2015. http:// timss.bc.edu.

Mullis, Ina V. S., Michael O. Martin, John F. Olson, Debra R. Berger, Dana Milne, Gabrielle M. Stanco, eds. *TIMSS 2007 Encyclopedia: A Guide to Mathematics and Science Education Around the World.* Vols. 1–2. Chestnut Hill, Mass.: TIMSS & PIRLS International Study Center, Boston College, 2008b. Accessed November 20, 2015. http://timss.bc.edu/timss2007 /encyclopedia.html

Mullis, Ina V. S., Michael O. Martin, Graham J. Ruddock, Christine Y. O'Sullivan, and Corinna Preuschoff. *TIMSS 2011 Assessment Frameworks*. Chestnut Hill, Mass.: TIMSS & PIRLS International Study Center, Boston College, 2009. Accessed November 20, 2015. http://timssandpirls.bc.edu/timss2011/downloads/TIMSS2011_Frameworks.pdf

National Assessment Governing Board (NAGB). *Mathematics Framework for the 2011 National Assessment of Educational Progress*. Washington, D.C.: NAGB, 2010. Accessed January 1, 2015. https://www.nagb.org/content/nagb/assets/documents/publications /frameworks/mathematics/2011-mathematics-framework.pdf

———. *Mathematics Framework for the 2015 National Assessment of Educational Progress*. Washington, D.C.: NAGB, 2014. Accessed September 7, 2015. https://www.nagb.org/content /nagb/assets/documents/publications/frameworks/mathematics/2015-mathematics -framework.pdf

———. *Frequently Asked Questions*. Washington, D.C.: NAGB, 2015. Accessed September 7, 2015. https://www.nagb.org/toolbar/faqs.html#naep

National Association of Charter School Authorizers (NACSA). *Overview of the State of Charter Authorizing: 2014*. Chicago: NACSA, 2014.

National Center for Education Information (NCEI). *Alternative Teacher Certification: A State-by-State Analysis, 2010*. Washington, D.C.: NCEI, 2010.

National Center for Education Statistics (NCES). *The Nation's Report Card: Mathematics 2009*. Washington, D.C.: U.S. Department of Education, NCES, 2009a. Accessed December 9, 2015. http://nces.ed.gov/nationsreportcard/pdf/main2009/2010451.pdf

———. *The Nation's Report Card: Grade 12 Reading and Mathematics State Snapshot Reports 2009*. Washington, D.C.: U.S. Department of Education, NCES, 2009b. Accessed December 9, 2015. http://nces.ed.gov/nationsreportcard/pdf/main2009/2010451.pdf

———. *The Nation's Report Card: Mathematics 2011*. Washington, D.C.: U.S. Department of Education, U.S. Department of Education, NCES, 2011. Accessed December 9, 2015. http://nces.ed.gov/nationsreportcard/pdf/main2011/2012458.pdf

———. *Interpreting NAEP Long-Term Trend Results*. Washington, D.C.: U.S. Department of Education, NCES, 2013a. Accessed September 12, 2015. https://nces.ed.gov /nationsreportcard/ltt/interpreting_results.aspx

———. *The Nation's Report Card: A First Look: Mathematics and Reading 2013*. Washington, D.C.: U.S. Department of Education, NCES, 2013b. Accessed December 9, 2015. http:// nces.ed.gov/pubsearch/pubsinfo.asp?pubid=2014451

———. *The Nation's Report Card: A First Look: 2013 Mathematics and Reading Trial Urban District Assessment*. NCES 2014-466. Washington, D.C.: U.S. Department of Education, NCES, 2013c. Accessed September 7, 2015. http://nces.ed.gov/nationsreportcard/subject /publications/main2013/pdf/2014466.pdf

———. *NAEP Mathematics 2013 Grade 12 State Snapshot Reports*. Washington, D.C.: U.S. Department of Education, NCES, 2013d. Accessed December 9, 2015. https://nces.ed.gov /pubsearch/pubsinfo.asp?pubid=2014082

———. *Trends in Academic Progress 2012*. Washington, D.C.: U.S. Department of Education, NCES, 2013e. Accessed December 9, 2015. http://nces.ed.gov/nationsreportcard /subject/publications/main2012/pdf/2013456.pdf

———. "Digest of Education Statistics: List of 2014 Tables and Figures." *Digest of Education Statistics*. Washington, D.C.: U.S. Department of Education, NCES, 2015a. Accessed December 17, 2015. https://nces.ed.gov/programs/digest/current_tables.asp

————. *Fast Facts: Charter Schools.* Washington, D.C.: U.S. Department of Education, NCES, 2015b. Accessed December 17, 2015. https://nces.ed.gov/fastfacts/display.asp?id=30

————. *NAEP Data Explorer.* Washington, D.C.: U.S. Department of Education, NCES, 2015c. Accessed December 9, 2015. https://nces.ed.gov/nationsreportcard/tdw/database/data_tool.asp

————. *The Nation's Report Card: A First Look: Mathematics and Reading 2015.* Washington, D.C.: U.S. Department of Education, NCES, 2015d. Accessed December 9, 2015. http://www.nationsreportcard.gov/reading_math_2015/#?grade=4

————. *School and Student Participation Rates in NAEP Mathematics at Grade 8, by State/Jurisdiction: 2015.* Washington, D.C.: U.S. Department of Education, NCES, 2016. Accessed March 31, 2016.

National Commission on Excellence in Education. *A Nation at Risk: The Imperative For Educational Reform—A Report to the Nation and the Secretary of Education.* Washington, D.C.: U.S. Department of Education, 1983. Accessed August 17, 2015. http://www2.ed.gov/pubs/NatAtRisk/

National Commission on Mathematics and Science Teaching for the 21st Century (Glenn Commission). *Before It's Too Late: A Report to the Nation from the National Commission on Mathematics and Science Teaching for the 21st Century.* Washington, D.C.: U.S. Department of Education, 2000. Accessed August 17, 2015. http://files.eric.ed.gov/fulltext/ED441705.pdf

National Consortium of Secondary STEM Schools (NCSSSMST). *History and Founders.* Chevy Chase, Md.: NCSSSMST, 2015. Accessed December 19, 2015. http://ncsss.org/about/history-and-founders

National Council of Teachers of Mathematics (NCTM). *An Agenda for Action: Recommendations for School Mathematics of the 1980s.* Reston, Va.: NCTM, 1980. Accessed August 17, 2015. http://www.nctm.org/flipbooks/standards/agendaforaction/index.html

————. *Curriculum and Evaluation Standards for School Mathematics.* Reston, Va.: NCTM, 1989.

————. *Professional Standards for Teaching Mathematics.* Reston, Va.: NCTM, 1991.

————. *Assessment Standards for School Mathematics.* Reston, Va.: NCTM, 1995.

————. *Principles and Standards for School Mathematics.* Reston, Va.: NCTM, 2000.

————. *Curriculum Focal Points for Prekindergarten through Grade 8 Mathematics: A Quest for Coherence.* Reston, Va.: NCTM, 2006.

————. *Mathematics Teaching Today: Improving Practice, Improving Student Learning.* Reston, Va.: NCTM, 2007.

————. *Focus in High School Mathematics: Reasoning and Sense Making.* Reston, Va.: NCTM, 2009.

————. *Making It Happen: A Guide to Interpreting and Implementing Common Core State Standards for Mathematics.* Reston, Va.: NCTM, 2011.

————. *Principles to Actions: Ensuring Mathematical Success for All.* Reston, Va.: NCTM, 2014.

————. *Supporting the Common Core State Standards for Mathematics.* NCTM Position Statement. Reston, Va.: NCTM, 2013. Accessed October 28, 2015. http://www.nctm.org

/Standards-and-Positions/Position-Statements/Supporting-the-Common-Core-State
-Standards-for-Mathematics/

National Governors Association Center for Best Practices and Council of Chief State School Officers (NGA Center and CCSSO). *Common Core State Standards for Mathematics. Common Core State Standards (College- and Career-Readiness Standards and K–12 Standards in English Language Arts and Math).* Washington, D.C.: NGA Center and CCSSO, 2010a. Accessed August 17, 2015. http://www.corestandards.org/Math/

————. *Appendix A: Common Core State Standards for Mathematics.* Washington, D.C.: NGA Center and CCSSO, 2010b. Accessed August 17, 2015. http://www.corestandards.org/assets /CCSSI_Mathematics_Appendix_A.pdf

National Institutes of Health. *Summer Institute for Training in Biostatistics.* Bethesda, Md.: National Institutes of Health, Office of Biostatistics Research, 2015. Accessed December 19, 2015. http://www.nhlbi.nih.gov/research/training/summer-institute-biostatistics-t15

National Mathematics Advisory Panel. *Foundations for Success: The Final Report of the National Mathematics Advisory Panel.* Washington, D.C.: U.S. Department of Education, 2008. Accessed August 17, 2015. https://www2.ed.gov/about/bdscomm/list/mathpanel/report/final -report.pdf

National Research Council. *Successful K-12 STEM Education: Identifying Effective Approaches in Science, Technology, Engineering, and Mathematics.* Committee on Highly Successful Science Programs for K–12 Science Education; Board on Science Education and Board on Testing and Assessment. Washington, D.C.: National Academies Press, 2011. Accessed August 17, 2015. http://www.nap.edu/catalog/13158/successful-k-12-stem-education -identifying-effective-approaches-in-science

————. *Fueling Innovation and Discovery: The Mathematical Sciences in the 21st Century.* Committee on the Mathematical Sciences in 2025; Board on Mathematical Sciences and Their Applications; Division on Engineering and Physical Sciences. Washington, D.C.: National Academies Press, 2012. Accessed August 17, 2015. http://www.nap.edu/catalog .php?record_id=13373

————. *The Mathematical Sciences in 2025.* Committee on the Mathematical Sciences in 2025; Board on Mathematical Sciences and Their Applications; Division on Engineering and Physical Sciences. Washington, D.C.: National Academies Press, 2013a. Accessed August 17, 2015. http://www.nap.edu/catalog/15269/the-mathematical-sciences-in-2025

————. *Monitoring Progress toward Successful K–12 STEM Education: A Nation Advancing?* Committee on the Evaluation Framework for Successful K–12 STEM Education; Board on Science Education and Board on Testing and Assessment; Division of Behavioral and Social Sciences and Education. Washington, D.C.: National Academies Press, 2013b. Accessed August 17, 2015. http://www.nap.edu/catalog/13158/successful-k-12-stem-education -identifying-effective-approaches-in-science

National Science Board (NSB). *Educating Americans for the 21st Century.* Commission on Precollege Education in Mathematics, Science, and Technology. Washington, D.C.: National Science Board, National Science Foundation, 1983. Accessed December 5, 2015. https:// ia802600.us.archive.org/19/items/ERIC_ED233913/ERIC_ED233913.pdf

————. *Science and Engineering Indicators 2014.* NSB 14-01. Arlington Va.: National Science Foundation, 2014. Accessed September 2, 2015. http://www.nsf.gov/statistics/seind14 /content/etc/nsb1401.pdf

National Science Foundation (NSF). *NSF REU Programs.* Arlington, Va.: National Science Foundation, 2015. Accessed December 22, 2015. www.nsf.gov/crssprgm/reu/

No Child Left Behind Act of 2001 (NCLB). Public Law 107-110. 107th Congress. Accessed September 1, 2015. http://www2.ed.gov/policy/elsec/leg/esea02/107-110.pdf

Nord, Christine, Shep Roey, Robert Perkins, Marsha Lyons, Nita Lemanski, Janis Brown, and Jason Schuknecht. *The Nation's Report Card: America's High School Graduates—Results of the 2009 NAEP High School Transcript Study.* Washington, D.C.: National Center for Educational Statistics, 2011. Accessed December 13, 2015. http://nces.ed.gov/nationsreportcard /pdf/studies/2011462.pdf

Organisation for Economic Cooperation and Development (OECD), *Learning for Tomorrow's World: First Results from PISA 2003.* Paris: PISA, OECD Publishing, 2004a. Accessed August 17, 2015. http://www.parcconline.org http://www.oecd.org/education/school /programmeforinternationalstudentassessmentpisa/34002216.pdf

———. *Problem Solving for Tomorrow's World: First Measures of Cross Curricular Competencies from PISA 2003.* Paris: PISA, OECD Publishing, 2004b. Accessed August 17, 2015. http://www.oecd.org/edu/school/programmeforinternationalstudentassessmentpisa /34009000.pdf

———. *PISA 2006 Science Competencies for Tomorrow's World, Volume 1 Analysis.* Paris: PISA, OECD Publishing, 2007. Accessed August 17, 2015. http://www.oecd.org/pisa /pisaproducts/39703267.pdf

———. *PISA 2009 Results: What Students Know and Can Do: Student Performance in Reading, Mathematics and Science.* Vol. 1. Paris: PISA, OECD Publishing, 2010. Accessed August 17, 2015. http://www.oecd.org/pisa/pisaproducts/48852548.pdf

———. *PISA 2012 Assessment and Analytical Framework: Mathematics, Reading, Science, Problem Solving and Financial Literacy,* Paris: OECD Publishing, 2013a. Accessed November 20, 2015. http://www.oecd.org/pisa/pisaproducts/PISA%202012 %20framework%20e-book_final.pdf

———. *PISA 2012 Results: What Students Know and Can Do: Student Performance in Mathematics, Reading and Science,* vol. I. Paris: PISA, OECD Publishing, 2013b. Accessed August 17, 2015. http://www.oecd.org/pisa/keyfindings/pisa-2012-results-volume-I.pdf

———. *PISA 2012 Results: Creative Problem Solving. Students' Skills in Tackling Real-Life Problems,* vol. 5. Paris: PISA, OECD Publishing, 2014. Accessed August 17, 2015. http:// www.meb.gov.tr/earged/oecd/PISA2012%20(vol%205)--eBook%20(eng)-FINAL.pdf

———. *PISA FAQ.* Paris: OECD Directorate for Education, 2016. Accessed April 11, 2016. https://www.oecd.org/pisa/aboutpisa/pisafaq.htm

Partnership for Assessment of Readiness for College and Careers (PARCC). States and Assessments. PARCC, 2015a. Accessed September 7, 2015. http://www.parcconline.org

———. *The Next Generation of Assessment.* PARCC, 2015b. Accessed August 17, 2015. http:// www.parcconline.org

Peak, Lois, and other NCES staff. *Pursuing Excellence: A Study of U.S. Fourth-Grade Mathematics and Science Achievement in International Context.* Washington, D.C.: National Center for Education Statistics, Institute of Education Sciences, U.S. Department of Education, 1995a. Accessed November 22, 2015. http://nces.ed.gov/timss/results95.asp

———. *Pursuing Excellence: A Study of U.S. Eighth-Grade Mathematics and Science Teaching, Learning, Curriculum, and Achievement in International Context: Initial Findings from the Third International Mathematics and Science Study.* Washington, D.C.: National Center for Education Statistics, Institute of Education Sciences, U.S. Department of Education, 1995b. Accessed November 22, 2015. http://nces.ed.gov/timss/results95.asp

Peck, Roxy, Chris Olsen, and Jay L. Devore. *Introduction to Statistics and Data Analysis.* 4th ed. Independence, Ky.: Cengage Learning, 2012.

Perie, Marianne, Wendy Grigg, and Gloria Dion, *The Nation's Report Card: Mathematics 2005.* Washington, D.C.: U.S. Department of Education, National Center for Education Statistics, 2005. Accessed December 9, 2015. http://nces.ed.gov/nationsreportcard/pdf /main2005/2006453.pdf

Perie, Marianne, Rebecca Moran, and Anthony D. Lutkus. *NAEP 2004 Trends in Academic Progress: Three Decades of Student Performance in Reading and Mathematics.* Washington, D.C.: National Center for Education Statistics, 2005. Accessed September 7, 2015. http://nces.ed.gov/nationsreportcard/pdf/main2005/2005464.pdf

President's Council of Advisors on Science and Technology. *Engage to Excel: Producing One Million Additional College Graduates with Degrees in Science, Technology, Engineering, and Mathematics.* Washington, D.C.: The White House, 2012. Accessed August 17, 2015. https://www.whitehouse.gov/sites/default/files/microsites/ostp/pcast-executive-report -final_2-13-12.pdf

Provasnik, Stephen, David Kastberg, David Ferraro, Nita Lemanski, Stephen Roey, and Frank Jenkins. *Highlights from TIMSS 2011: Mathematics and Science Achievement of U.S. Fourth- and Eighth-Grade Students in an International Context.* NCES 2013-009. Washington, D.C.: National Center for Education Statistics, Institute of Education Sciences, U.S. Department of Education, 2012. Accessed November 20, 2015. http:// nces.ed.gov/pubs2013/2013009_1.pdf

Rampey, Bobby D., Gloria S. Dion, and Patricia L. Donahue, *NAEP 2008 Trends in Academic Progress.* Washington, D.C.: U.S. Department of Education, National Center for Education Statistics, 2009. Accessed December 9, 2015. http://nces.ed.gov/nationsreportcard/pdf /main2008/2009479.pdf

Ray, Brian D. "Home Schooling: The Ameliorator of Negative Influences on Learning?" *Peabody Journal of Education* 75, nos. 1–2 (2000): 71–106.

————. *Research Facts on Homeschooling.* Salem, Ore.: National Home Education Research Institute (NHERI), 2015. Accessed December 3, 2015. http://www.nheri.org/research/research -facts-on-homeschooling.html

Reese, Clyde M., Karen E. Miller, John Mazzeo, and John A. Dossey. *NAEP 1996 Mathematics Report Card for the Nation and the States.* Princeton, N.J.: Educational Testing Service and National Center for Education Statistics, 1997. Accessed December 9, 2015. http:// nces.ed.gov/nationsreportcard//pdf/main1996/97488.pdf

Resnick, Robert M., and Glenn Sanislo. *The Complete K–12 Report: Market Facts & Segment Analyses 2015.* Rockaway Park, N.Y.: Education Market Research/Simba Information, 2015.

Reys, Barbara, ed. *The Intended Mathematics Curriculum as Represented in State-Level Curriculum Standards: Consensus or Confusion?* Charlotte, N.C.: Information Age, 2006.

Robitaille, David F., and Robert A. Garden, eds. *The IEA Study of Mathematics II: Contexts and Outcomes of School Mathematics.* Oxford, UK: Pergamon Press, 1989.

Romberg, Thomas A. *School Mathematics: Options for the 1990s.* Washington, D.C.: U.S. Department of Education, 1984.

Ross Mathematics Program at the Ohio State University. *Introduction.* Columbus, Ohio: Ross Program, Ohio State University Department of Mathematics, 2015. Accessed December 19, 2015. http://u.osu.edu/rossmath/

Schmidt, William. "Researchers Find Systemic Problem in U.S. Mathematics and Science Education." *New Education.* East Lansing, Mich.: College of Education, Michigan State University, fall 2000. Accessed December 5, 2015. http://www.educ.msu.edu/neweducator/fall00/timss.htm

————. "At the Precipice: The Story of Mathematics Education in the United States." *Peabody Journal of Education* 87, no. 1 (2012): 133–56, 2012.

Shettle, Carolyn, Shep Roey, Joy Mordica, Robert Perkins, Christine Nord, Jelena Teodorovic, Janis Brown, Marsha Lyons, Chris Averett, and David Kastberg. *America's High School Graduates: Results from the 2005 NAEP High School Transcript Study.* Washington, D.C.: National Center for Education Statistics, 2007. Accessed January 1, 2016. https://nces.ed.gov/nationsreportcard/pdf/studies/2007467.pdf

Shober, Arnold B. "ESEA Reauthorization Continues a Long Federal Retreat from American Classrooms." *The Brown Center Chalkboard.* Washington, D.C.: Brookings Institution, December 8, 2015.

Smarter Balanced Assessment Consortium (SBAC). *Understand the Individual Student Report.* SBAC, 2015. Accessed August 17, 2015. http://smarterbalanced.org

Smith, Adrienne A. *2012 National Survey of Science and Mathematics Education: Status of High School Mathematics.* Chapel Hill, N.C.: Horizon Research, 2013.

Smith, John (Jack) P. *Variability Is the Rule: A Companion Analysis of K–8 State Mathematics Standards.* Charlotte, N.C.: Information Age, 2010.

Snook, Kathleen. "Mathematics and Mathematical Sciences in 2010: The How of What Graduates Should Know, 2010." In *Discussion Papers about Mathematics and the Mathematical Sciences in 2010: What Should Students Know?* edited by the Committee on the Undergraduate Program in Mathematics, pp. 87–93. Washington, D.C.: Mathematical Association of America, 2004. Accessed August 17, 2015. http://www.maa.org/cupm/ about.html

Snyder, Thomas B., and Sally A. Dillow. *Digest of Education Statistics 2013.* Washington, D.C.: National Center for Education Statistics, 2015. Accessed August 1, 2015. http://nces.ed.gov

Society for Industrial and Applied Mathematics (SIAM). *SIAM Report on Mathematics in Industry 2012.* Philadelphia: SIAM, 2012. Accessed August 17, 2015. https://www.siam.org/reports/mii/2012/index.php

Society for Industrial and Applied Mathematics and Consortium for Mathematics and Its Applications (SIAM and COMAP). *Guidelines for Assessment and Instruction in Mathematics Modeling Education.* Philadelphia: SIAM; Bedford, Mass.: COMAP, 2016.

Stark, Patricia, and Amber M. Noel. *Trends in High School Dropout and Completion Rates in the United States: 1972–2012.* NCES 2015-015. Washington, D.C.: U.S. Department of Education. National Center for Education Statistics, 2015. Accessed August 1, 2015. https://nces.ed.gov/pubsearch/pubsinfo.asp?pubid=2015015

Starnes, Daren, Josh Tabor, Dan Yates, and David S. Moore. *The Practice of Statistics.* 5th ed. New York: Bedford, Freeman, and Worth, 2015.

Stetser, Marie C., and Robert Stillwell. *Public High School Four-Year On-Time Graduation Rates and Event Dropout Rates: School Years 2010–11 and 2011–12—First Look.* NCES 2014-391. Washington, D.C.: National Center for Education Statistics, 2014. Accessed October 16, 2015. http://nces.ed.gov/pubs2014/2014391.pdf

Stewart, James. *Calculus.* 8th ed. Independence, Ky.: Cengage Learning, 2016.

Takahira, Sayuri, Patrick Gonzales, Mary Frase, and Laura Hersh Salganik. *Pursuing Excellence: A Study of U.S. Twelfth-Grade Mathematics and Science Achievement in International*

Context (1995). Washington, D.C.: National Center for Education Statistics, Institute of Education Sciences, U.S. Department of Education, 1995. Accessed November 22, 2015. http://nces.ed.gov/timss/results95.asp

TIMSS & PIRLS International Study Center. *About TIMSS and PIRLS*. Chestnut Hill, Mass.: TIMSS & PIRLS International Study Center, Boston College, 2015. Accessed November 20, 2015. http://timssandpirls.bc.edu/home/pdf/TP_About.pdf

Travers, Kenneth J., and Ian Westbury, eds. *The IEA Study of Mathematics I: Analysis of Mathematics Curricula*. Oxford, UK: Pergamon Press, 1989.

UICSM Staff. "The University of Illinois School Mathematics Program." *School Review* 65 (Winter 1957): 457–65.

U.S. Department of Commerce. *American Community Survey*, Washington, D.C.: U.S. Census Bureau, American Community Survey Office, 2007.

———. *American Community Survey*, Washington, D.C.: U.S. Census Bureau, American Community Survey Office, 2012.

———. *Current Population Survey* (CPS), Washington, D.C.: U.S. Census Bureau, American Community Survey Office, 2014.

U.S. Department of Education. *No Child Left Behind*. Washington, D.C.: U.S. Department of Education, 2008. Accessed August 4, 2015. http://www2.ed.gov/nclb/landing.jhtml

———. *State Regulation of Private Schools*. Washington, D.C.: U.S. Department of Education, Office of Innovation and Improvement, Office of Non-Public Education, 2009.

UTeach Institute. *UTeach and UTeach Expansion: Data Through Spring 2015*, Austin, Tex.: College of Natural Sciences, University of Texas at Austin, 2015. Accessed November 22, 2015. https://institute.uteach.utexas.edu/sites/institute.uteach.utexas.edu/files/uteach-institute-press-kit-june-2015.zip

National Council of Teachers of Mathematics

The National Council of Teachers of Mathematics is the public voice of mathematics education, providing vision, leadership, and professional development to support teachers in ensuring mathematics learning of the highest quality for all students. With 70,000 members and more than 200 Affiliates, NCTM is the world's largest organization dedicated to improving mathematics education in prekindergarten through grade 12. The Council's *Principles and Standards for School Mathematics* includes guidelines for excellence in mathematics education and issues a call for all students to engage in more challenging mathematics. Its *Curriculum Focal Points for Prekindergarten through Grade 8 Mathematics* identifies the most important mathematical topics for each grade level. *Focus in High School Mathematics: Reasoning and Sense Making* advocates practical changes to the high school mathematics curriculum to refocus learning on reasoning and sense making. NCTM's *Principles to Actions: Ensuring Mathematical Success for All* describes the principles and actions, including specific research-informed teaching practices, that are essential for a high-quality mathematics education for all students. NCTM is dedicated to ongoing dialogue and constructive discussion with all stakeholders about what is best for our nation's students. For more information on NCTM or the most up-to-date listing of NCTM resources on topics of interest to mathematics educators, as well as information on membership benefits, conferences, and institutes, visit the NCTM website at www.nctm.org, contact NCTM at inquiries @nctm.org, call (703) 620-9840, or follow NCTM on Twitter or Facebook.

United States National Commission on Mathematics Instruction

The U.S. National Commission on Mathematics Instruction (USNC/MI) is a committee of the U.S. National Academy of Sciences. The roles of the USNC/MI are to facilitate U.S. participation in the activities of the International Commission on Mathematical Instruction and to engage the U.S. mathematics education community through the National Council of Teachers of Mathematics, the Conference Board of the Mathematical Sciences, and the National Research Council to advance mathematics education in the United States and throughout the world. Support to the USNC/MI is provided by the National Science Foundation. Further information is available by contacting usncmi@nas.edu.